T0133348

V.V.Blednykh, P.G. Svechnikov

Theory of a tillage wedge and its applications

2013

UDC 631.31

LBC 40.722

B 688

BLEDNYKH V.V., SVECHNIKOV P.G.

Theory of a tillage wedge and its applications [Text]

Monograph / V.V.Blednykh, P.G. Svechnikov. 2013. 90 p.

Reviewers

S.G. Mudarisov - Doctor of Technical Sciences, Professor, Head of the Department "Agricultural machinery" FSBEI HPE "Bashkir State Agrarian University"

V.A. Smelik - Doctor of Technical Sciences, Professor, Head of the Department "Technological Processes and machines for crop production" FSBEI HPE "St. Petersburg State Agrarian University"

Bibliographic information published by the Deutsche Nationalbibliothek

The Deutsche Nationalbibliothek lists this publication in the Deutsche Nationalbibliografie; detailed bibliographic data are available in the Internet at http://dnb.d-nb.de .

ISBN 978-3-8325-3550-6

Logos Verlag Berlin GmbH

Comeniushof, Gubener Str. 47,

10243 Berlin

Tel.: +49 (0)30 42 85 10 90

Fax: +49 (0)30 42 85 10 92

INTERNET: http://www.logos-verlag.de

CONTENTS

INTRODUCTION

The founder of the agricultural mechanics V.P. Goryachkin [23] noticed that "... the theory of every instrument must answer two questions:

1) what form should have a working part of a tool for the most advanced quality of work;

2) what should be the size and location of all tool components (active and inactive) for the most comfortable operation with the smallest possible expenditure of effort .... ". That is, first, the theory "... studies the configuration change of the processed material and the movement of the particles by the working parts of a tool (for example, the particles of earth by a blade surface) from the point of view of the quality of work, regardless of the magnitude of the forces and necessary costs of effort ...." and then "... determines the amount of work depending on one or the other form of all parts of the tool and their relative position ...."[23].

Thus, first, a working surface is produced, the results of the work of this surface meeting the specifications, and then the acting force and energy consumption for the production of this work are determined.

Therefore, "... the question of the form of a tool is a major one, and the effort expenditure − minor, since every instrument is assessed primarily on the quality rather than the quantity of work ..." [23].

As you know, the diversity of the working bodies of tillers can be studied on the example of a wedge with the corresponding spatial angles. We take the view, according to which the main task of tillage is to change the structure of the treated soil for favorable germination and development of crop plants.

The deformation process by a dihedral wedge according to the scheme proposed by V.P. Goryachkin [23, 24] is as follows: when driving a wedge is initially compacted, the movement of its particles is perpendicular to the work surface. The character of fracture thus depends on the wedge angle α and physical and mechanical properties of the soil. Medium-textured loams with humidity 18 ... 20% are deformed with a shift at an angle to the direction of the movement of the soil wedge .P.M. Vasilenko distinguishes three periods of the interaction of soil with a wedge [18]:

1) the crushing of a soil layer;

2) the increase of the stress state with the formation of individual cracks at an angle ψ to the direction of the wedge movement:

$$\psi = \frac{\pi}{2} - \frac{\varepsilon_1 + \varphi + \rho}{2} \ ,$$

where ε_1 – the angle of the setting of the working face of the wedge to the bottom of a furrow;

φ – the angle of the soil friction on the surface of the wedge;

ρ – the angle of the internal friction of the soil;

3) the destruction of the soil.

G.N. Sineokov [52, 53], considers that the motion of a skew wedge can be represented as composed of two interlaced elementary displacements: perpendicular to the blade of the wedge that causes the deformation and cutting of the layer and parallel to the blade, which does not cause the deformation in the plane perpendicular to the blade.

In the above context, the soil treatment should be carried out so as to obtain the desired quality of treatment, reduce energy consumption and increase the production of machinery. The optimal quality of soil crumbling the required turnover of the layer etc. can be anticipated, causing the layer of soil to move along a predetermined trajectory on the reversible-moldboard surface of a moldboard plow or on the share of a moldboardless tool.

In this context the in-depth research of the interaction of the working bodies with the soil in order to modernize and design of the working bodies of tillage tools that allow you to receive the set agronomic data of soil tillage with minimal costs is necessary.

In the theory of a wedge there are still many gaps, and the working bodies for soil tillage so far have been created on the base of heuristic guesswork. In this paper, we try to answer some of the unsolved theoretical issues, putting forward scientific hypotheses proven by us experimentally.

1. MAIN CHARACTERISTICS OF THE WEDGE

Since the days of the founder of agricultural mechanics in Russia V.P. Goryachkin in the wedge theory it is accepted to distinguish between a dihedral wedge and a trihedral one.

A dihedral wedge is understood as a wedge, when the work process is performed in the plane.

Usually such a scheme is considered (Fig. 1): the wedge ABC with an angle α moves in the soil at a depth of a. For the purpose of clarity, the influence of friction force while moving the soil on the wedge in the diagram sometimes the friction angle of the soil movement on the surface of the wedge is referred to as φ. When $\alpha + \varphi = 90°$, the soil movement on the wedge is impossible.

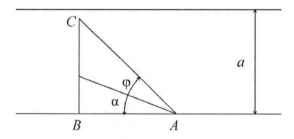

Fig. 1. Dihedral wedge

In case of a trihedral wedge the interaction process with soil is performed in space. A trihedral wedge is characterized by two major (constructional) angles — ε and γ. The angle ε defines the installation of the wedge plane to the bottom (surface) of a furrow, and the angle γ between the plane ABC wedge and the plane coinciding with the movement direction — OBC (Fig. 2). The vast majority of the working bodies of tillers is a sort of a dihedral or trihedral wedge. In the theory of a triangular wedge ancillary angles α and β are considered. It is accepted that the angle β characterizes the ability of a wedge to the turnover of a soil layer and the angle α —

the ability to destroy (crumble) the layer. The angles α and β are easily obtained from the following geometric transformations:

$$\mathrm{tg}\beta = \frac{OC}{OA}, \quad \mathrm{tg}\gamma = \frac{OA}{OB}, \quad \mathrm{tg}\alpha = \frac{OC}{OB},$$

that is

$$\mathrm{tg}\gamma\,\mathrm{tg}\beta = \frac{OA}{OB}\frac{OC}{OA} = \frac{OC}{OB} = \mathrm{tg}\alpha, \quad \mathrm{tg}\alpha = \mathrm{tg}\gamma \cdot \mathrm{tg}\beta \qquad (1)$$

$$\mathrm{tg}\varepsilon = \frac{OC}{OD}; \quad \sin\gamma = \frac{OD}{OB}; \quad \mathrm{tg}\varepsilon\sin\gamma = \frac{OC}{OD}\frac{OD}{OB} = \frac{OC}{OB} = \mathrm{tg}\alpha$$

or

$$\mathrm{tg}\alpha = \mathrm{tg}\varepsilon \cdot \sin\gamma \qquad (2)$$

Equating (1) and (2), we obtain another relation:

$$\mathrm{tg}\beta = \mathrm{tg}\varepsilon \cdot \cos\gamma \qquad (3)$$

If you know the technological (constructional) angles γ and ε (Fig. 2), it is enough to set one any side of the wedge to determine the other sides.

Example
It is given: γ, ε, $OB = X$. We define the other parameters of the wedge:

$$OD = X\sin\gamma,$$

$$\sin\gamma = \frac{OD}{OB}; \quad \text{tg}\varepsilon = \frac{OC}{OD}, \text{ т. е. } OC = OD\text{tg}\varepsilon = X\sin\gamma\,\text{tg}\varepsilon\,; \quad OA = X\text{tg}\gamma\,.$$

It is given: γ, ε, OC = Z, etc.

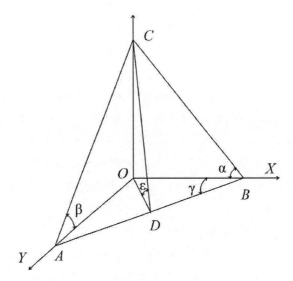

Fig. 2. Trihedral wedge

2. INTERACTION OF A FLAT DIHEDRAL WEDGE WITH SOIL

In the study of tillers designers are usually interested in the following manifestations of the interaction of working bodied with soil:

− the deformation (destruction) of soil;

− the movement of soil and bodies in the soil;

− the forces that arise during the interaction of working bodies with soil;

− the impact of the shape of the working body on the listed manifestations of the interaction.

The forces that arise in the interaction of working bodies with soil should be overcome. The knowledge of these forces enables to determine the value of tool traction, the type of a traction tool, the energy and the economy of the given tool. To create a reliable and durable machine, you need to learn how to work with the forces. With proper account of all the forces the possibility to quickly create good working bodies is very high.

We can suppose that the wedge BAO with an angle at apex α is moving in the soil at a depth of a (Fig. 3). On the horizontal movement of the wedge in the direction of V the soil answers the reaction N. If there is no friction on the wedge, the wedge interacts with the soil along the normal N. If the reaction of N is directed as shown in Figure 3 a, the force N will be the reaction of the wedge to the action of soil. If the force N is directed as shown in Figure 3b, then N will be the reaction of soil to the action of a wedge.

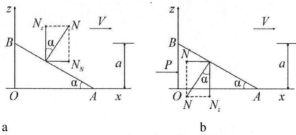

a   b

Fig. 3. Scheme of the direction options of a normal reaction between the wedge and the soil in the absence of friction: a − the direction of the normal reaction of the wedge; b − the direction of the normal soil reaction.

Since the friction along the wedge in the soil movement is always present, while moving the soil along the wedge the interaction of the wedge with the soil will be in the direction of the force R, the resultant of N and F (Fig. 4, 5), wherein:

$$F = N \cdot \mathrm{tg}\varphi,$$

where φ –the angle of friction of the soil movement on the wedge.

$$R = \frac{N}{\cos\varphi}.$$

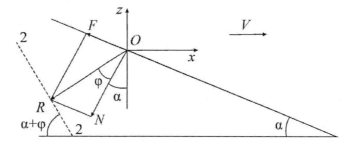

Fig. 4. Soil reaction when driving on a surface with friction (rough surface)

The reaction of the wedge R is a force under the influence of which the soil is destroyed. Under the influence of the vertical component of the reaction wedge R_z the soil is moving upward. The horizontal component of the soil reaction Rx is a major component of the force of traction resistance of the wedge P, while the vertical component of the soil reaction R_z presses the wedge to the axis ox (the bottom of the furrow).

The condition of the wedge movement will be like this (Fig. 5):

$$P = R_x + R_z \cdot \mathrm{tg}\varphi_1,$$

where $\mathrm{tg}\varphi 1$ − the coefficient of the friction of the wedge sole OA on the ground; $\varphi 1$ –the angle of the sliding of the wedge sole on the ground.

11

But

$$R_x = R\sin(\alpha + \varphi);$$

$$R_z = R\cos(\alpha + \varphi).$$

Therefore

$$P = R\left[\sin(\alpha + \varphi) + \cos(\alpha + \varphi)\mathrm{tg}\varphi_1\right]. \tag{4}$$

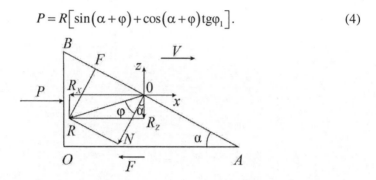

Fig. 5. Stationary process of moving the dihedral wedge in the soil

If the material and structural design of the sole of the wedge are the same with the material and structural design of the working surface of the wedge, then $\varphi_1 = \varphi$, that is,

$$P = \frac{N}{\cos\varphi}\left[\sin(\alpha + \varphi) + \cos(\alpha + \varphi)\mathrm{tg}\varphi\right]. \tag{5}$$

2.1. Destruction of the soil by a dihedral wedge

Unfortunately, in the world little attention is paid to the study of variation of soil properties while mechanical processing, the necessary instruments for the measurement of physical quantities characterizing the soil are absent. To determine the possibility of the quality crumbling of soil at the lowest cost it is necessary to know the resistance of the soil to the destruction under different conditions of

12

interaction of working bodies with the soil. There have been no generally accepted models of the destruction (strain) of the soil so far. A significant number of well-known models of soil degradation use the laws of the destruction of an elastic body. The essence of these methods: the voltage at the internal points of the solid body, arising under the impact of the forces, can be represented if mentally a portion of the body is allocated. This selected part of the body will be acted by the forces of the rest parts of the body, the strain on the surface of the selected part will occur. These stresses are subject to the following conditions:

− the force applied to the selected volume, should be equal to zero in the rest state;

− the forces applied to the selected volume should be equal to the product of the mass of the selected volume by its acceleration in the case of movement.

In addition, similar conditions for the moments of these forces must be carried out. Numerous attempts to apply the methods of investigation of elastic materials for the analysis of the crumbling (destruction) of the soil have not yielded practical results yet, as the relationship between the force acting on the soil , and the deformation of the soil is the function of a soil condition.

We recall the basic properties of an elastic body:

− between the load on the body and its deformation there is a linear relationship;

− the elastic properties of the body at all points are the same and do not depend on direction;

− an elastic body is always unbroken (before deformation and after deformation);

− a deformed elastic body after the removal of the external load returns to its original state. In this case all the work expended on the deformation returns too.

Any soil that is regularly exposed to treatment, does not have this set of properties. Therefore, the use of a mechanical and mathematical unit of the destruction of elastic bodies in the classical form while designing the working bodies is ineffective. It should be noted that depending on the texture, moisture and degree of compaction, soils can sometimes have separate fragment properties of an elastic body. Direct criteria of strength while resisting the soil to destruction are defined for many types of soils under various humidity [1, 2, 22, 23, 27, 28, 33, 34, 37, 54]:

$\sigma_{cж}$ – compressive strength;

σ_p – tensile strength;

τ – tensile shear strength..

The lowest limit of the soil strength is observed in tension (σ_p), the greatest limit is observed in compression ($\sigma_{cж}$), the middle value – in shear (τ). Professor V. A. Zhilkin obtained the following values for the tensile strength of the clay black soil (Table 1) [26].

Table 1

Temporary stretch, compression and shear resistance of clay black soil

Tensile		Compression		Shift	
Soil humidity, %	$\sigma_p \cdot 10^3$, mPa	Soil humidity, %	$\sigma_{cж} \cdot 10^3$, mPa	Soil humidity, %	$\tau \cdot 10^3$, mPa
21–23	6,18	12–16	108,00	15–17	12,21
23–25	2,25	19–22	98,00	20–24	9,86
26–28	5,00	22–24	65,00	–	–

According to M.Ya. Zhuk [27], for the clay black soil at the humidity $W_a = 20 \ldots$ 23% $\sigma_p = 6-7$kPa; $\sigma_{cж} = 90 - 98$ kPa; $\tau = 9-10$ kPa. It should be specifically defined and distinguished the strength of the matted connected soil from which you can cut a monolith and expose it to the tests and the strength of individual structural elements of the soil, usually loosely connected.

Indicators of the limit strength of soil degradation can only serve as a reference for evaluating the actual deformation of the soil in the interaction with the working bodies of the tools. Usually, while studying the interaction of working bodies with the soil, you can not specify in advance the form of deformation which the soil was undergone. Let us consider one of the possible approaches to the study of the process of the soil destruction by the working bodies of the tools. The basic tenets of the approach are:

– the destruction of the soil and its components occurs when the active force exceeds the strength to resist fracture;

– the magnitude and direction of the force can be determined as a result of the analysis of the interaction of the working body with the ground;

– the force to resist the destruction of soil, can be very roughly defined on the basis of the mechanical composition of the soil and its moisture content at the time of processing.

The key elements of the process of the soil destruction by a dihedral wedge are shown in figure 6.

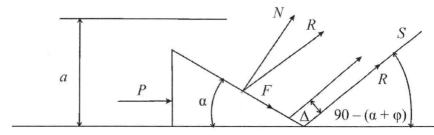

Fig. 6. The destruction of the soil by a dihedral wedge

When driving in the soil the wedge interacts with the soil along the normal, deflected by an angle of the friction of the soil movement on the wedge surface. In this direction the wedge moves the ground on the plane in an area:

$$S = \frac{ab}{\sin\left(90 - (\alpha + \varphi)\right)} \quad \text{или} \quad S = \frac{ab}{\cos(\alpha + \varphi)}, \qquad (6)$$

where S – the plane of the soil destruction;

a – the depth of the wedge stroke;

α – the angle of the dihedral wedge;

φ – the friction angle when sliding the soil on the wedge;

b –the width of the wedge.

To produce soil degradation in the plane S, the amount of the force R must be equal to

15

$$R = \frac{\mu ab}{\cos(\alpha + \varphi)}, \tag{7}$$

where μ – the coefficient of the clutch of the soil particles in the area of S.

Of course, with large assumptions one can regard this coefficient to be the limit soil resistance to the shear τ. The action of the force R can manifest only when on the wedge, the soil layer is formed with a thickness of an element of the latter, equal to Δ, i.e., a wedge is introduced into the soil without destroying it as long as on the wedge the layer element with the thickness Δ is not formed. The thickness of the layer element can be determined from the condition

$$R = \Delta b\sigma, \tag{8}$$

where σ – the limit normal stress on the layer element with the thickness Δ (this may be the compressive strength of the soil $\sigma_{cж}$).

Solving equations (8) and (7), we have:

$$\Delta = \frac{a\tau}{\sigma \cos(\alpha + \varphi)} \text{ или } = \Delta = \frac{\mu a}{\sigma \cos(\alpha + \varphi)}. \tag{9}$$

Figure 7 shows the dependence of the thickness of the layer element Δ on the angle of the wedge mounting to the furrow bottom defined by equation 9, provided $\sigma_{cж} = 10\tau$.

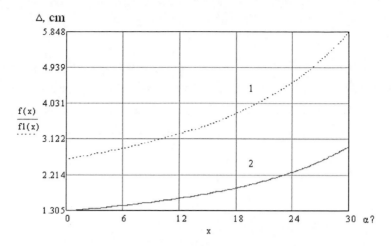

Fig. 7.Thickness of the soil layer elements according to the position of the wedge angle to the bottom of the furrow ($\sigma_{\text{ск}}$ = 10т; φ = 40°; 1 − a = 20 cm; 2 − a = 10 cm)

Figure 8 shows the value of the force R on the wedge, calculated according to equation 7.

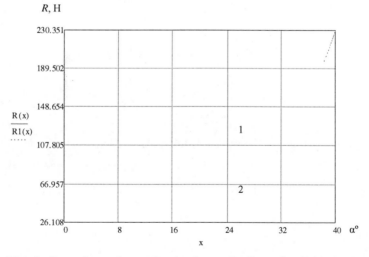

Fig. 8. Srength on the wedge in the separation of soil layer element with the thickness Δ (1 − the depth of the wedge stroke 0.2 m; 2 − the depth of the wedge stroke 0.1 m; μ = 2000 N/m2; b = 0,1 m)

17

2.2. Formation of a soil layer on a wedge

The layer of soil on the wedge – is constantly interacting particles of the soil, limited by the size of the wedge. The movement of the soil layer on the wedge is implemented at the expense of the layer interaction with the mainland soil. The process of the layer formation on the dihedral wedge is shown in figure 9. The soil layer element separated from the mainland with the thickness Δ enters the wedge to form a new layer element with the thickness Δ, etc.

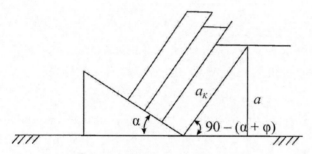

Fig. 9.Layer formation on the wedge

As a result, the layer with the thickness a_K i s formed on the wedge:

$$a_K = \frac{a}{\cos(\alpha + \varphi)} .$$
(10)

We have performed numerous experimental studies of the process of forming a layer on the wedge. The experiments confirmed the above scheme of the formation of the soil layer on the wedge from the layer elements with the thickness Δ (fig. 10).

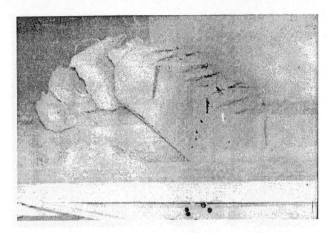

Fig. 10. Formation of the soil layer on the wedge (the soil – the sand with humidity 14%)

We have discovered that in the process of the layer formation on a flat wedge the layer with a bigger thickness than the depth of the working body stroke is formed. Initially, the experimental data were approximated with the dependence [38]:

$$\frac{a_K}{a} = \mathrm{tg}\alpha + \cos\alpha,$$

(11)

where a_K – the layer thickness of the wedge, m;

a – the depth of the wedge stroke, m;

α – the wedge setting angle to the bottom of the furrow.

Calculations show that dependence 11 coincides satisfactorily with dependence 10 (fig. 11).

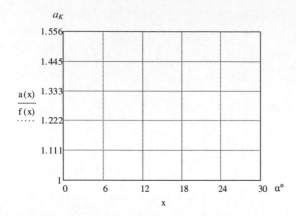

Fig. 11. Estimated thickness of the layer on the wedge, calculated from equations 10 and 11 (a = 1)

2.3. Movement of a soil layer on a dihedral wedge

Despite an apparent simplicity, unfortunately, so far there have been no mathematical equations by which it would be possible to calculate the contribution of the soil movement on the wedge in the draft resistance of the wedge, including the effect of the coefficient of the soil movement friction on the wedge on its draft resistance. We shall try to fill this gap. Let us consider the condition of the motion of a generated soil layer on *abcd* on the flat dihedral wedge with an angle α (Fig. 12). The length of the wedge plane is *l*, the width of the wedge plane is equal to *b*.

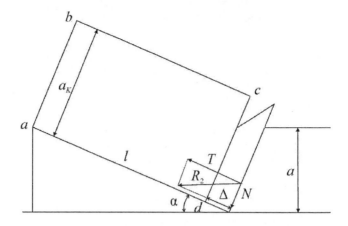

Fig. 12. Estimated pattern of the soil layer traffic *abcd* on the wedge with an angle α

The soil volume on the wedge:

$$V = a_K bl,$$

where a_K –the layer thickness on the wedge;
b, l – the length and width of the working plane of the wedge.
The weight of the soil on the wedge:

$$G = \gamma V,$$

where γ –the unit weight of the soil.

The normal component on the plane of the wedge from the strength of the soil weight:

$$G_N = G \sin \alpha = \gamma V \sin \alpha.$$

The resistance force to the soil layer movement on the wedge:

$$T = G_N f \,,$$ (12)

where f – the coefficient of the friction for the motion of the soil on the wedge.

The strength of T results from the interaction of the formed layer with the mainland soil (fig. 12).

While interacting the soil layer on the wedge with the mainland soil, the following relations take place:

$$T = R_2 \cos\alpha \,;$$

$$N = R_2 \sin\alpha \,,$$

where R_2 – the force, preventing the wedge movement in the soil (resistance force);

N – the force normal to the plane of the wedge.

The power o N increases the frictional force of the soil movement on the wedge on the value fN, so the equation (12) will be:

$$T = G_N f + Nf \,,$$

or

$$T = fG_N + fR_2 \sin\alpha \,,$$

or

$$R_2 \cos\alpha = fG_N + fR_2 \sin\alpha \,,$$

whence we find the force R_2:

$$R_2 = \frac{fG\sin\alpha}{\cos\alpha - f\sin\alpha},$$ (13)

i.e.

$$R_2 = \frac{f\gamma a_\kappa bl\sin\alpha}{\cos\alpha - f\sin\alpha}.$$ (14)

The thickness of the layer on the dihedral wedge with the angle α is defined by the equation (10), i.e. the final expression for determining the contribution of the soil movement on the wedge in draft resistance of the wedge will be

$$R_2 = \frac{f\gamma abl\sin\alpha}{\cos(\alpha + \varphi)(\cos\alpha - f\sin\alpha)}.$$ (15)

The resulting equation can answer all questions about the effects of the motion of the soil layer on the wedge on its draft resistance. When the given size of the wedge and the volumetric weight of the soil (γ = 2000 kg/m3) the layer weight on the wedge is 10−14 kg. The dependence of the traction resistance on the wedge setting angle to the bottom of the furrow and on the coefficient of the soil friction on the surface of the wedge, calculated according to the equation (15), is shown in figures 13 and 14, which show that the draft resistance of the wedge is comparable to the layer weight with a coefficient of the sliding soil friction on the wedge f = 0.5.

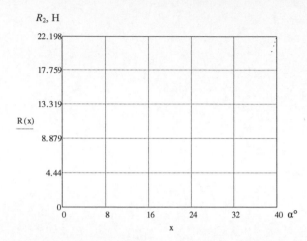

Fig. 13. Draft resistance of the wedge, depending on the setting angle of the wedge to the bottom of the furrow, caused by the movement of the soil layer on the wedge

$(F = 0.5; a = 0,1 \text{ m}; b = 0,2 \text{ m}; l = 0,3 \text{ m}; \gamma = 2000 \text{ kg/m3})$

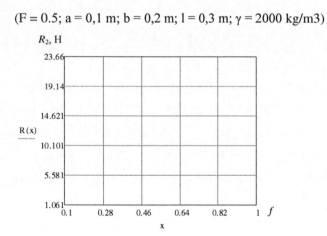

Fig. 14. Effect of the soil friction against the wedge surface on its tractive resistance $(\varepsilon = 30°; a = 0,1 \text{ m}; b = 0,2 \text{ m}; l = 0,3 \text{ m}; \gamma = 2000 \text{ kg/m3})$

2.4. Formation of a dihedral wedge during the movement of a harrow tooth in soil.

When moving a harrow tooth, the soil cannot move up and in sides, so a soil wedge is formed with an angle of internal friction of the given soil − μ. This wedge in work is like a dihedral wedge. Now the running soil can move along the soil wedge up and bring field weeds to the surface.

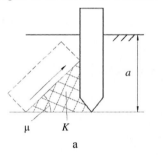 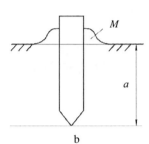

a   b

Fig. 15. Movement of a harrow tooth in the soil: a − the formation of the soil wedge K; b – the output of a micro layer to the surface M

2.5. Tractive resistance of a dihedral wedge

The strength of the draft resistance of the wedge P is determined by the force of the soil destruction R (fig. 6, equation (7)), the force of the wedge sole friction on the bottom of the furrow F2 (fig. 5) and the friction force of the soil movement on the wedge at the time of the destruction of F1 (still without power R2, expended to move the formed soil layer on the wedge with the length l):

$$P = P_\delta + F_1 + F_2,$$ (16)

where − P_p the draft resistance from the process of the soil destruction:

$$P_p = R\cos(90 - (\alpha + \varphi)) = R\sin(\alpha + \varphi);$$

F1 – the tractive resistance of friction on the wedge in the destruction of the soil;

$$F_1 = N\mathrm{tg}\varphi;$$

N – the normal force of pressure on the surface in the destruction of the mainland soil (fig. 12);

φ – the angle of the soil movement friction on the wedge:

$$N = \frac{R}{\cos\varphi}; \quad F_1 = \frac{fR}{\cos\varphi};$$

F2 – the tractive resistance from the force of the wedge sole friction on the furrow bottom:

$$F_2 = R_z\mathrm{tg}\varphi,$$

where R_z – the force pushing the wedge to the furrow bottom:

$$R_z = R\cos(\alpha + \varphi) + F_1\sin\alpha = R\left(\cos(\alpha + \varphi) + \frac{f\sin\alpha}{\cos\varphi}\right),$$

$$F_2 = fR\left(\cos(\alpha + \varphi) + \frac{f\sin\alpha}{\cos\varphi}\right).$$

So, we have:

$$P = P_\delta + F_1 + F_2 = R\left(\sin(\alpha + \varphi) + \frac{f}{\cos\varphi} + f\left(\cos(\alpha + \varphi) + \frac{f\sin\alpha}{\cos\varphi}\right)\right). \quad (17)$$

After substitution of R in the equation (17) we have:

$$P=\frac{\mu ab}{\cos(\alpha+\phi)}\left[\sin(\alpha+\phi)+\frac{f}{\cos\phi}+f\left(\cos(\alpha+\phi)+\frac{f\sin\alpha}{\cos\phi}\right)\right]. \qquad (18)$$

The influence of the movement of the soil layer on the draft resistance is given by the equation (15) and can always be taken into account. Figure 16 shows the total value and the components of the force of traction resistance, calculated from equations 7, 17 and 18.

The influence of the force of the sole friction against the furrow bottom on the draft wedge resistance is much smaller than the influence of the force of the friction against the wedge surface at the time of the soil destruction.

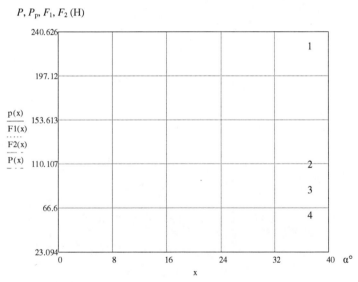

Fig. 16.Ttotal value and leaving forces of the traction resistance of the dihedral wedge: 1 − the total value of the force of traction resistance of the wedge; 2 − P_p − the tractive resistance by the soil degradation; 3 − F_1− the tractive resistance by the friction at the time of the soil destruction; 4 − F_2 − the tractive resistance by the wedge sole friction on the furrow bottom (taken in the calculations a = b = 0.1 m; φ = 30°; f = 0,58; μ = 4000 H/m2)

The considered model of the soil destruction is consistent with the physical processes of interaction of a wedge with soil, and the diagrams in figure 16 show that the calculated data agree quite well with the experimental data obtained by many authors in Russia and abroad (17, 25, 29, 31, 34 , 35, 51, 52, 53, 54, 56).

The convergence of the given models with the experimental data is much higher than the convergence of other models, including those based on elastic properties of the soil.

3. INTERACTION OF A TRIHEDRAL WEDGE WITH SOIL

The place of a trihedral wedge in tillers defines the particular qualities of a technological process performed by the wedge: the availability of plant residues in the soil (roots etc.), as well as the necessity, together with loosening, to shift the ground towards the direction of motion. In the study of the interaction of the triangular wedge with the soil the most important are the following indicators of the wedge:

- the trajectory of the soil on the wedge;
- the power characteristics of the wedge;
- the speed of the soil movement on the wedge.

In such a sequence these questions will be considered.

3.1. Trajectory of soil movement along a wedge

If from the top of the wedge C we drop a perpendicular to the blade AB (point D), and point D we connect to the origin of coordinates, the angle ODC will be an angle ε. ε − the setting angle of the plane of the wedge ABC to the furrow bottom. OD also is perpendicular to AB (fig. 17).

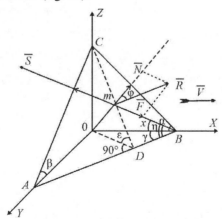

Fig. 17. Trajectory of the soil movement and the forces acting on a triangular wedge during its movement in the soil

From the origin of coordinates we draw a perpendicular to the plane of the wedge ABC. We obtain the point of intersection of the perpendicular with the plane of the wedge – point m. In the direction Om the reaction of interaction of the wedge with the soil $\bar{N}$ proceeds.

The soil moves along the trajectory S on the wedge, defined by an angle η. The angle η – is the angle between the path of the movement of the soil and the wedge edge AB. On the trajectory S the force of friction $F = N \cdot tg\varphi$ in the direction opposite to the motion proceeds. An angle OmD is direct ($\angle OmD = 90°$). Also the angles are direct: $\angle mDB = 90°$, $\angle ODB = 90°$. Such a representation of the mechanics of the process of movement makes it easy to determine the angle of η:

$$tg\eta = \frac{mD}{BD};\ tg\gamma = \frac{OD}{BD};\ \cos\varepsilon = \frac{mD}{OD};$$

$$tg\gamma\cos\varepsilon = \frac{OD}{BD}\frac{mD}{OD} = \frac{mD}{BD} = tg\eta;$$

$$tg\eta = tg\gamma \cdot \cos\varepsilon. \qquad\qquad (19$$

The angle η determines the trajectory of the soil movement.

Figure 18 shows the effect of the angles γ and ε on the trajectory of the soil movement on the wedge. In the literature the angle η is called **the angle of entry of the soil on the wedge.**

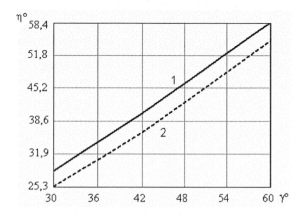

Fig. 18.Ttrajectory of the soil movement on a trihedral wedge: $1 - \varepsilon = 20°$; $2 - \varepsilon = 35°$

The knowledge of the angle η enables to determine the cutting angle of the soil by a triangular wedge χ. This angle lies in a plane OmB (fig. 17) and is measured by the angle between the direction of the movement of the wedge (axis x), and the trajectory of the soil movement S. The cutting angle is presented as one of the summarizing structural indicators of a trihedral wedge and is analogous to an angle α in the dihedral wedge. If you know the cutting angle χ, then all the equations for the dihedral wedge can be used with a certain error, substituting the angle α by the angle χ. We calculate the angle χ:

$$\sin\chi = \frac{Om}{OB}; \quad \sin\gamma = \frac{OD}{OB}; \quad \sin\varepsilon = \frac{Om}{OD},$$

$$\sin\gamma\sin\varepsilon = \frac{OD}{OB}\frac{Om}{OD} = \frac{Om}{OB} = \sin\chi,$$

$$\sin\chi = \sin\gamma \cdot \sin\varepsilon. \tag{20}$$

The meaning of the cutting angle χ depending on the installation angles of the triangular wedge γ and ε is presented in figure 19. Hence the cutting angle of a trihedral wedge is always less than any of the mounting angles.

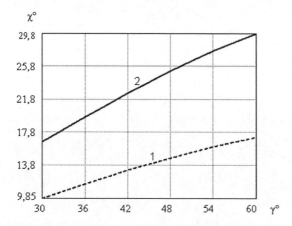

Fig. 19. Soil cutting angle of a triangular wedge: $1 - \varepsilon = 20°$; $2 - \varepsilon = 35°$

Let us estimate the draft resistance of the triangular wedge in a similar way to the calculations of the dihedral wedge, which angle at the apex is equal to the cutting angle χ (fig. 21).

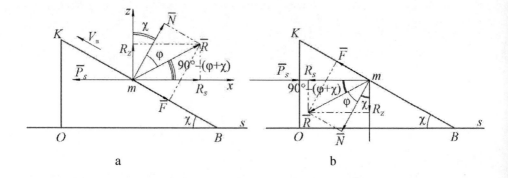

a   b

Fig. 20. Scheme for the determination of the force P: a – the direction of the reaction of the wedge; b – the direction of the soil reaction when driving a wedge in it

Figure 20 a shows a reaction $\overline{N}$, forces $\overline{R}$ and $\overline{F}$ have the same direction as in figure 3. To determine the resistance to the movement of the wedge P_S , it is necessary to show the reaction of the soil, which must be overcome by a wedge (fig. 20 b). While seemingly identical the images of the wedge interaction with the soil in the diagrams a and b (Fig. 20), they have a different physical meaning, which is necessary to take into account when solving practical problems.

So:

$$R = \frac{N}{\cos\varphi}; \; R_S = R\cos\left[90° - (\chi+\varphi)\right] = \frac{N}{\cos\varphi}\sin(\chi+\varphi);$$

$$R_Z = R\cos(\chi+\varphi);$$

$$P_S = R_S + R_Z \cdot \mathrm{tg}\varphi_1.$$

We have the following expression for the draft resistance of the triangular wedge:

$$P_S = R\left[\sin(\chi+\varphi) + \cos(\chi+\varphi)\cdot \mathrm{tg}\varphi_1\right]. \tag{21}$$

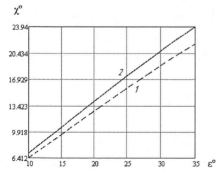

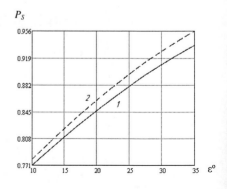

Fig. 21. Angles of cut and draft resistance of the triangular wedge under the following variable values: $\varphi = \varphi1 = 20\,°$; $N = 1$ $1 - \gamma = 40\,°$; $2 - \gamma = 45\,°$

3.2. Power characteristics of a trihedral wedge

The triangular wedge causes the presence of components of traction resistance in all axes of coordinates. The reliable experimental results to determine the components of the traction resistance of the body of a moldboard plow on the axes are obtained by G.N. Sineokov [52, 53].

We define the components of the force $\bar{R}$ of the triangular wedge on the axes of Cartesian coordinates. For brevity, we omit the standard arguments and present only the derivation of the basic relationships:

$$\bar{R} = \bar{N} + \bar{F}; \ \ \bar{R} = \left(R_x + N_x; \ R_y + N_y; \ R_z + N_z \right).$$

Coordinates of the corners of the wedge (fig. 22): $A\left(0; \ y_a; 0 \right)$; $B\left(x_b; 0; 0 \right)$; $C\left(0; 0; z_c \right)$; $\overline{AB} = \left(x_b; - y_a; 0 \right)$; $\overline{AC} = \left(0; - y_a; z_c \right)$.

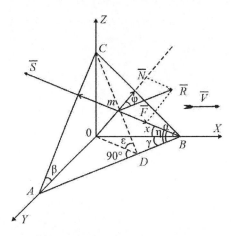

Fig. 22. Forces acting on a triangular wedge as it moves into the soil

We denote the unit vector $\overline{N_0}$, normal to the plane ABC.
Then

$$\overline{N} = B \cdot \overline{N_0},$$

where V − any positive number:

$$\overline{N_0} = \frac{\overline{AB} \cdot \overline{AC}}{\left| \overline{AB} \cdot \overline{AC} \right|};$$

$$\overline{AB} \cdot \overline{AC} = \begin{vmatrix} \overline{i} & \overline{j} & \overline{k} \\ x_b & -y_a & 0 \\ 0 & -y_a & z_c \end{vmatrix} = -y_a z_c \overline{i} - x_b z_c \overline{j} - x_b y_a \overline{k} = \left(-y_a z_c ; -x_b z_c ; -x_b y_a \right);$$

$$\left| \overline{AB} \cdot \overline{AC} \right| = \sqrt{y_a^2 z_c^2 + x_b^2 z_c^2 - x_b^2 y_a^2} = d.$$

$$N_x = \frac{|\overline{N}| \cdot y_a z_c}{d}; \quad N_y = \frac{|\overline{N}| \cdot x_b z_c}{d}; \quad N_z = \frac{|\overline{N}| \cdot x_b y_a}{d}.$$

Using the same mathematical formalism, we define the force $\overline{F}$ components along the coordinate axes:

$$\overline{F} = q \cdot \overline{F}_0,$$

where $\overline{F}$ – the unit vector;

$$q = |\overline{N}| \cdot tg\varphi,$$

where φ – the angle of the soil friction on the surface of the wedge in motion:

$$\overline{F}_0 = \frac{m\overline{B}}{|m\overline{B}|}; \quad m\overline{B} = (x_b - x_m; \, y_b - y_m; \, z_b - z_m).$$

Since $y_b = 0; \, z_b = 0$, then $m\overline{B} = (x_b - x_m; \, -y_m; \, -z_m)$.

$$|m\overline{B}| = \sqrt{(x_b - x_m)^2 + (y_b - y_m)^2 + (z_b - z_m)^2} = e;$$

$$F_x = |\overline{N}| \cdot tg\varphi \cdot \frac{x_b - x_m}{e}; \quad F_y = |\overline{N}| \cdot tg\varphi \cdot \frac{(-y_m)}{e}; \quad F_z = |\overline{N}| \cdot tg\varphi \cdot \frac{(-z_m)}{e}.$$

To complete the process of finding $\overline{F}(F_x; \, F_y; \, F_z)$, we find the coordinates of the point $m(x_m; \, y_m; \, z_m)$. The coordinates of the point m can be determined based on the following theorem: the m point lies in the plane ABC and only when the vectors $\overline{Am}$,

$\overline{AB}$ и $\overline{AC}$ are co-planar. The condition of the three vectors to be co-planar is the equality to 0 of the determinant of the third order, made up of their coordinates:

$$\begin{vmatrix} x_m - x_a & y_m - y_a & z_m - z_a \\ x_b - x_a & y_b - y_a & z_b - z_a \\ x_c - x_a & y_c - y_a & z_c - z_a \end{vmatrix} = 0 ;$$

$$\begin{vmatrix} x_m - 0 & y_m - y_a & z_m - 0 \\ x_b - 0 & 0 - y_a & 0 - 0 \\ 0 - 0 & 0 - y_a & z_c - 0 \end{vmatrix} = -x_m y_a z_c - x_b y_a z_m - x_b y_m z_c + x_b y_a z_c = 0 ;$$

$$\frac{x_m}{N_x} = \frac{y_m}{N_y} = \frac{z_m}{N_m} ; \ \rightarrow \ x_m = \frac{N_x}{N_y} y_m = \frac{y_a}{x_b} y_m, \ \ y_m = \frac{N_y}{N_z} z_m = \frac{z_c}{y_a} z_m ;$$

$$x_m y_a z_c + x_b y_a z_m + x_b y_m z_c = x_b y_a z_c ;$$

$$\begin{cases} x_m - \dfrac{y_a}{y_b} y_m = 0, \\[3mm] y_m - \dfrac{z_c}{y_a} z_m = 0, \\[3mm] \left(x_b y_a + \dfrac{z_c^2}{y_a} x_b + \dfrac{z_c^2}{x_b} y_a \right) \cdot z_m = x_b y_a z_c ; \end{cases}$$

$$\frac{d^2 \cdot z_m}{x_b y_a} = x_b y_a z_c ;$$

$$z_m = \frac{x_b^2 \cdot y_a^2 \cdot z_c}{d^2} ; \quad y_m = \frac{x_b^2 \cdot z_c^2 \cdot y_a}{d^2} ; \quad x_m = \frac{y_a^2 \cdot z_c^2 \cdot x_b}{d^2} .$$

The calculations allow us to determine all the components of the forces on the axes $\overline{N}\left(N_x; N_y; N_z\right)$ и $\overline{F}\left(F_x; F_y; F_z\right)$ and calculate the components of the force $\overline{R}$ along the coordinate axes:

$$
\left.
\begin{aligned}
R_x &= N_x + F_x \\
R_y &= N_y + F_y \\
R_z &= N_z + F_z
\end{aligned}
\right\} .
\tag{22}
$$

In the first approximation, the force of traction resistance of the triangular wedge ABC during its even motion in the direction $\overline{V}$ is:

$$
P_x = R_x + R_z \mathrm{tg}\varphi_1 + R_y \mathrm{tg}\varphi_2,
\tag{23}
$$

where φ1 – the friction angle of the lower edge of the wedge on the bottom of the furrow;

φ2 – the friction angle of the side edge of the wedge on the wall of the furrow.

If φ = φ1 = φ2, then

$$
P_x = R_x + \mathrm{tg}\varphi\left(R_z + R_y\right).
\tag{23a}
$$

The values P_x, and the relationships R_y / P_x and R_z / P_x for N = 1 and different angles ε and γ are presented in figures 23-25.

P_x, force is a force that must be overcome in order to cause the wedge to move. This force is called **the force of traction resistance**. While designing it is important to pay attention to the equation (23), as the materials which are used to make the work surface of the wedge (φ), the support surface in the plane yox (φ1) and in the plane zox (φ2), play an important role in the distribution of efforts to coordinate axes at the same normal reactions $\overline{N}$ (fig. 25).

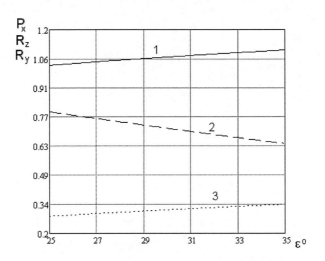

Fig. 23. Horizontal component of the draft resistance of the triangular wedge P_x and the relations R_y/P_x и R_z/P_x ($N = 1$, $\gamma = 40°$, $\varphi = \varphi_1 = \varphi_2 = 20°$): $1 - P_x$; $2 - R_z/P_x$; $3 - R_y/P_x$

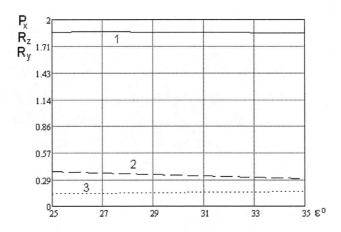

Fig. 24. Components of the traction resistance of the triangular wedge
($N = 1$, $\gamma = 40°$, $\varphi = \varphi_1 = \varphi_2 = 40°$): $1 - P_x$; $2 - R_z/P_x$; $3 - R_y/P_x$

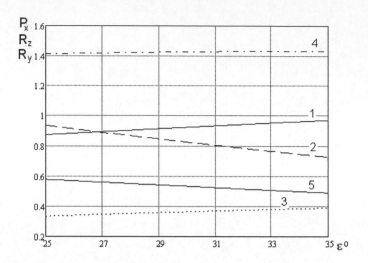

Fig. 25. Components of the traction resistance of the triangular wedge

($N = 1$, $\gamma = 40°$, $\varphi_1 = \varphi_2 = 20°$): $1 - P_x$ при $\varphi_2 = 10°$; $2 - R_z/P_x$ при $\varphi_2 = 10°$; $3 - R_y/P_x$;

$4 - P_x$ при $\varphi_2 = 40°$; $5 - R_z/P_x$ при $\varphi_2 = 40°$

From the figures 21, 23 and 25, it follows that the use of the equation (21) for determining the traction resistance of the triangular wedge gives a greater error and it can be used only for a very rough estimate of the traction resistance of the wedge (fig. 21).

For the design of the working bodies, the technological processes which correspond to agro-technical requirements, it is necessary to know the path and speed of the soil movement along the coordinate axes. For this it is necessary to determine the direction cosines of the trajectory of the soil movement S:

$$S_x = S\cos\alpha, \ V_x = V\cos\alpha; \ S_y = S\cos\beta, \ V_y = V\cos\beta; \ S_z = S\cos\gamma, \ V_z = V\cos\gamma,$$

where α – the angle of the trajectory of the soil movement with the axis X;

β – the angle of the trajectory of the soil movement with the axis Y;

γ – the angle of the trajectory of the soil movement with the axis Z.

We write the equation of the path S, using the fact that S passes through the points B and F (fig. 26).

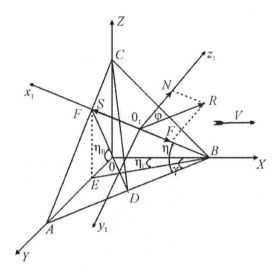

Fig. 26. Determination of the direction cosines of the trajectory S

$$\frac{x - x_1}{x_2 - x_1} = \frac{y - y_1}{y_2 - y_1} = \frac{z - z_1}{z_2 - z_1}.$$ (24)

The canonical form of the equation (24):

$$\frac{x - x_1}{l} = \frac{y - y_1}{m} = \frac{z - z_1}{n}.$$ (24a)

The coordinates of the point B are known: 0; OB; 0 (0; x_1, 0). The coordinates of the point F (see fig. 26) we find in the relations:

$$y_2 = OE = OBtg\eta_{\bar{A}};$$

41

$$z_2 = FE = OE\operatorname{tg}\eta_{\text{Å}}\operatorname{tg}\eta_{\text{В}};$$

$$x_2 = 0.$$

After substituting in the equation (24 a) the found values of the coordinates of the points B and F we get:

$$l = x_2 - x_1 = -OB; \quad m = y - y = OB\operatorname{tg}\eta_{\text{Å}}; \quad n = z_2 - z_1 = OB\operatorname{tg}\eta_{\text{Å}}\operatorname{tg}\eta_{\text{Å}}.$$

We express $\eta_{\text{Г}}$ и $\eta_{\text{В}}$ through corners η, γ, ε:

$$\operatorname{tg}\eta_{\text{Å}} = \frac{\operatorname{tg}\gamma - \cos\varepsilon\cdot\operatorname{tg}\eta}{1 + \cos\varepsilon\cdot\operatorname{tg}\gamma\cdot\operatorname{tg}\eta}; \qquad (25)$$

$$\operatorname{tg}\eta_{\text{Å}} = \frac{\operatorname{tg}\eta\cdot\sin\varepsilon}{\cos\gamma(\operatorname{tg}\gamma - \cos\varepsilon\cdot\operatorname{tg}\eta)}; \qquad (26)$$

$$\operatorname{tg}\eta_{\text{Å}}\cdot\operatorname{tg}\eta_{\text{Å}} = \frac{\operatorname{tg}\eta\cdot\sin\varepsilon}{\cos\gamma(\operatorname{tg}\gamma - \cos\varepsilon\cdot\operatorname{tg}\eta)}\cdot\frac{\operatorname{tg}\gamma - \cos\varepsilon\cdot\operatorname{tg}\eta}{1 + \cos\varepsilon\cdot\operatorname{tg}\gamma\cdot\operatorname{tg}\eta};$$

$$\operatorname{tg}\eta_{\text{Å}}\cdot\operatorname{tg}\eta_{\text{Å}} = \frac{\operatorname{tg}\eta\cdot\sin\varepsilon}{\cos\gamma(1 + \cos\varepsilon\cdot\operatorname{tg}\gamma\cdot\operatorname{tg}\eta)}. \qquad (27)$$

We find the direction cosines of the vector $\vec{S}$:

$$\cos\alpha_{\vec{s}} = \frac{m}{\sqrt{l^2 + m^2 + n^2}} = -(\cos\eta\cdot\cos\gamma + \cos\varepsilon\cdot\sin\gamma\cdot\sin\eta); \qquad (28)$$

$$\cos\beta_{\bar{s}} = \frac{l}{\sqrt{l^2 + m^2 + n^2}} = \cos\eta \cdot \sin\gamma - \cos\varepsilon \cdot \cos\gamma \cdot \sin\eta; \qquad (29)$$

$$\cos\gamma_{\bar{s}} = \frac{n}{\sqrt{l^2 + m^2 + n^2}} = \sin\varepsilon \cdot \sin\eta. \qquad (30)$$

3.3. Speed of soil movement along a wedge

We have determined the direction cosines of the trajectory of the layer movement along the trihedral wedge S (equations 28-30). This makes it possible to determine the velocity of the layer movement on Cartesian axes:

$$V_{kx} = V_k \cdot \cos\alpha_s,$$
$$V_{ky} = V_k \cdot \cos\beta_s, \qquad (31)$$
$$V_{kz} = V_k \cdot \cos\gamma_s,$$

where $\cos\alpha_s$, $\cos\beta_s$, $\cos\gamma_s$ – the direction cosines of the trajectory of the layer motion;

V_k – the velocity of the soil movement along the wedge (the path S);

x – the axis, which coincides with the direction of motion (fig. 26).

The rate of the soil on the wedge (fig. 27) is defined from the condition of the continuity of the movement of the soil:

$$aV = a_K V_k,$$

as

$$a_K = \frac{a}{\cos(\chi + \varphi)},$$

then

$$V_k = V\cos(\chi + \varphi),$$

where χ –the cutting angle f the soil by the triangular wedge (fig. 28).

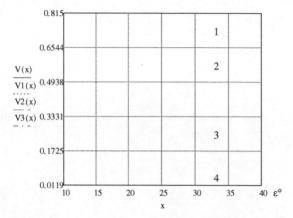

Fig. 27. Velocity of the soil movement along the wedge and the velocity components V_k along the coordinate axes (the wedge movement speed $V = 1$, $\gamma = 40°$, $\varphi = 30°$): 1 – V_k; 2 – V_x; 3 – V_z; 4 – V_y

The values of the cutting angle for the considered values of the angles γ and ε are shown in figure 28.

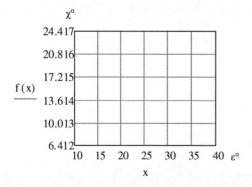

Fig. 28. Cutting angle of the soil χ ($\gamma = 40°$)

44

3.4. Blade of a trihedral wedge

Weeding feet are one of the main working bodies. They tend to work in heavy weed infestation – the roots, rhizomes, plant residues. During the feet movement in the soil the residues are cut by the blade (if it is very sharp), or slide along the blade until the exit from the blade. If the sliding velocity on the blade of roots is insufficient, the rhizomes are collected on the blade, the draft resistance of weeding feet increases sharply and they go up. Let us consider the process.

Let the blade of the weeding foot AB moves in the direction of V, in this case (fig. 29):

b – the width of the blade coverage , m;

γ – the angle of blade setting AB to the direction of motion;

Z_0 – the number of roots and other plant residues at the depth of the blade movement over an area of 1 m², pcs/m².

If the tiller moves across the field at the speed V (m/s), then the following number of roots enters the blade at a time, pcs/sec:

$$z_1 = Z_0 \cdot bV .$$

We assume that all arriving roots at the blade are not cut, and they slide on the blade until the descent from it.

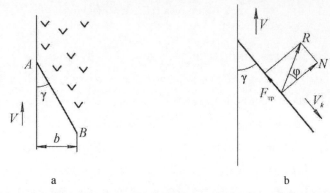

a   b

Fig. 29.Iinteraction of the blade of the wedge with the roots: a − the wedge movement in the direction $\vec{V}$; b − the forces of interaction between the blade of the wedge with the roots

In order not to break the technological process of weeding (cultivation), one need such a speed of the root movement along the blade, where the number of roots coming down from the blade is equal to the number of roots received for any period of time. If the speed of the roots on the blade is denoted by V_k , then the number of coming down roots in a unit of time is:

$$z_2 = z_1 \cdot V_k, \text{ items/sec,}$$

where z_1 – the number of roots per unit of the blade length, pcs/ m.

The condition of a stationary state of the technological process of the foot blade operation:

$$z_1 = z_2;$$

$$
\begin{cases}
Z_0 \cdot bV = z_1 \cdot V_k, \\
b = \dfrac{z_1 \cdot V_k}{Z_0 \cdot V}, \\
\dfrac{V_k}{V} = b \dfrac{Z_0}{z_1}.
\end{cases}
\tag{32}
$$

46

We define the sliding velocity of the roots on the blade, with which the condition of steady-state process is performed.

The interaction of the blade with the roots proceeds along the normal N (fig. 29). If $\gamma < 90 - \varphi$, where φ – the angle of the root friction when driving on the blade, then the roots are moving along the blade, and the friction force $F_{\text{тр}}$,directed to the side opposite the movement.

Power R, the resulting of forces N and $F_{\text{тр}}$, determines the trajectory of the root movement in the field plane (such a trajectory is convincingly demonstrated on the device by V.A. Zheligovskiy).

If from the sock of the blade AB (the beginning of movement) we draw a straight line having the friction angle φ with a normal, then the point of intersection of a straight line in the direction R with the movement direction of the end of the blade B-B1 would mean coming the root off the blade (fig. 30).

Moving at a certain speed V_k along the blade, the root during the time t will go the path from the beginning (p. A) to the end of the blade (p. B (B1)). During the same time t, moving along with the cultivator with speed V, the blade will go the path AA1, i.e.

$$AB = V_k \cdot t; \ AA_1 = V \cdot t,$$

whence

$$\frac{AB}{V_k} = \frac{AA_1}{V} \ \text{или} \ V_k = V \cdot \frac{AB}{AA_1}. \tag{33}$$

We find the value of AA1 by the law of sines (fig. 30):

$$\frac{AA_1}{\sin(90+\varphi)} = \frac{A_1 B_1}{\sin\left[90-(\gamma+\varphi)\right]},$$

whence

$$AA_1 = AB\frac{\cos\varphi}{\cos(\gamma+\varphi)}, \text{ i.e. } AB = A_1B_1.$$

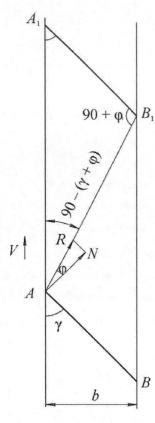

Fig. 30. Scheme for determining the slip rate of the roots along the blade

After substituting the value of AA_1 in the equation (33) we have:

$$\left.\begin{aligned} V_k &= V\frac{\cos(\gamma+\varphi)}{\cos\varphi}, \\ \frac{V_k}{V} &= \frac{\cos(\gamma+\varphi)}{\cos\varphi}. \end{aligned}\right\} \qquad (34)$$

According to calculations, V_k is highly dependent on the angle of the foot blade setting to the direction of the motion γ (fig. 31). Therefore, in practice in the weeding feet the angle γ = 28 ... 30 ° (for the wing shares 2γ = 60 and 65 °). For these values of the angle γ the root sliding velocity on the blade roots is 50 ... 70% of the movement speed of the cultivator.

Solving together the equations (34) и (32), we find:

$$b = \frac{z_1 \cdot \cos(\gamma+\varphi)}{Z_0 \cdot \cos\varphi}; \qquad (35)$$

$$z_1 = bZ_0\frac{\cos\varphi}{\cos(\gamma+\varphi)}. \qquad (36)$$

The number of roots per a length unit of the blade (z_1, pcs/m) increases with increasing foot coverage and the infestation of a processed field (fig. 32), and for certain values of the coverage width of the blade and the infestation it leads to a loss of foot efficiency: it goes up due to overhanging roots. Therefore the industry produces weeding feet with different feet coverage widths (b = 0.08-0.25 m for one-way clutches) and (b = 0.22 ... 0.4 m for wing shares). In the treatment of fields with little infestation feet with greater coverage width are used (fig. 33).

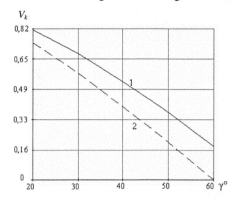

Fig. 31. Velocity of the roots when sliding along the blade of the wedge V_k during the machine speed V = 1: 1 − φ = 20 °; 2 − φ = 30 °

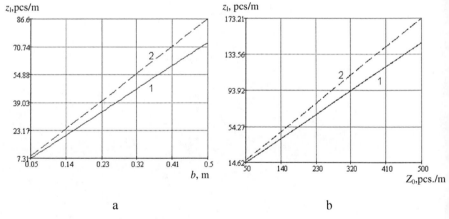

a   b

Fig. 32. Number of roots per a length unit of the blade: a − depending on the blade coverage width (b, m): Z_0 = 100 pcs/ m², 1 − γ = 30 °, φ = 20 °; 2 − γ = 30 °, φ = 30 °; b − depending on the field infestation (Z_0): b = 0.2 m; 1 − γ = 30 °, φ = 20 °; 2 − γ = 30 °, φ = 30 °

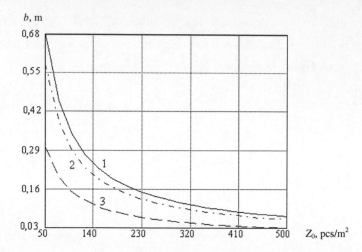

Fig. 33. Workable coverage width of a one side foot depending on the weed infestation of a field (Z_0) and the type of roots (φ): $1 - \gamma = 30$ °, $\varphi = 20$ ° $[Z_1] = 50$ pcs/ m; $2 - \gamma = 30$ °, $\varphi = 30$ ° $[Z_1] = 50$ pcs. / M $3 - \gamma = 45$ °, $\varphi = 30$ ° $[Z_1] = 50$ pcs/m

4. COVERAGE WIDTH OF CULTIVATOR FEET AND ITS OPTIMIZATION

Coverage width of cultivator feet cannot be determined arbitrarily. In order the foot to be workable it is necessary: plant residues in the soil, during the interaction with the blade grips should cut or slide along the blade until the exit from the blade. In section 3.4, the condition was obtained, under which the residues go off the blade (eq. (35)). For a stubble share, this equation becomes:

$$\frac{B}{2} = b \le \frac{[z_1] \cdot \cos(\gamma + \varphi)}{Z_0 \cdot \cos \varphi}, \tag{37}$$

where φ – the angle of friction of plant residues on the metal blade; $[Z_1]$ – the maximum allowable number of roots (rhizomes) per meter of the blade width under which the direct contact of the soil with the blade is not observed; $[Z_1] = 20 \ldots 150$ pcs/ m for a root diameter of 1.0 ... 2, 5 mm, and the higher value corresponds to sandy soils, the less one – to clay soils;

Z_0 – the infestation of soil with roots, pcs/m ^{2}.

The analysis of the equation (37) shows that the specific values of the angle γ correspond to the maximum values B (fig. 35). It is very important for creating less metal-working and stronger bodies.

As it can be seen from figures 35 and 36, the designer has the option of a larger selection of different combinations of the coverage width of the working body and the mounting angle of the coulter blade to the direction of motion. The selection area can be substantially narrowed, bearing in mind that the length of the working body L (fig. 34), the width of its capture (b) and the angle γ are related by

$$b = L \cdot \sin \gamma,$$
$$B = 2b.$$

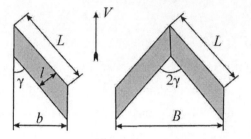

Fig. 34. Main design parameters of wing shares of cultivators:
L –the length of the plowshare blade; b – the coverage width of the plowshare; γ
– the mounting angle of the share to the direction of motion; l – the width of the
ploughshare steel

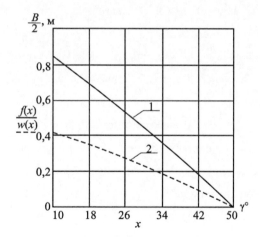

Fig. 35. Utmost coverage width of the working body blade depending on
the angle of its installation, and the field infestation: $1 - Z_0 = 20$ pcs / m ^{2}; $2 - Z_0$
$= 40$ pcs/ m ^{2};$[Z_1] = 20$ pcs/ m; $\varphi = 40$ °

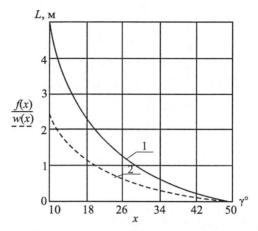

Fig. 36. Length of the plowshare, depending on the angle of installation, and the field infestation:

$$1 - Z_0 = 20 \text{ pcs/m}^2; \; 2 - Z_0 = 40 \text{ pcs/}; \; [z_1] = 20 \text{ pcs/m}^2; \; \varphi = 40°$$

For any design the indicator of metal content of technology is very important. In this case, as an indicator of metal content of a technological process, implemented by a cultivator foot, can act a value of metal mass on a coverage width meter:

$$\mu = \frac{G}{b},$$

where μ – the metal content designs, kg/m;

   G – the working body weight, kg ;

   b – the coverage width of a working body, m.

Assuming that the width and thickness of the ploughshare steel of the working body are constant, the length of the working body will unequivocally characterize its weight ($G \equiv L$).

The indicator of the working body metal content in this case will be (fig. 37):

$$\mu = \frac{L}{b} = \frac{l}{\sin\gamma}. \tag{38}$$

From the expression (38) it follows that in order to reduce a metal consumption indicator it is advantageous to have the maximum possible angle γ. With the reduction of the working body coverage width the number of racks on which the working body is mounted on the frame of the tool increases. Depending on the technology functions of a tool, the forces occurring at the rack can be very large, whereby the rack weight with the mounted hardware can exceed the mass of the working body.

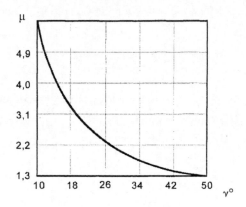

Fig. 37. Dependence of the metal content of the technological process μ (in conventional units) from the angle of the share installation to the wall of a furrow γ

This situation leads to the need to solve the problem of optimizing the coverage width of a working body for the tool having the coverage width B by the metal content criterion:

$f(B)$ = the mass of metal racks + the mass of metal feet $\rightarrow$ min;

$$f(B) = g_{\text{ст}} + g_{\text{л}} \rightarrow \text{min}, \tag{39}$$

provided (restrictions) the agronomic requirements for the crop residues slip and loosening parameters are performed (equations 32, 34).

We define the components of the objective function:

$$g_{\text{по}} = n \cdot g_{1c},$$

where g_{1c} – the mass of a rack $g_{1c} = f(b)$;

n –the number of racks.

Similar to:

$$g_{\text{ё}} = n \cdot g_{1\text{ё}},$$

where $g_{1\text{л}}$ – the mass of a foot; $g_{1\text{л}} = f(b)$;

b – the width of the coverage of a working body;

$$n = \frac{B}{b},$$

where B –the coverage width of a tool.

So, we have the problem:

$$n \cdot \left(g_{1c} + g_{1\text{ё}} \right) \rightarrow \min.$$

We introduce an indicator of metal content of a technological process

$$\mu = \frac{rack\ mass + feet\ mass}{B} \quad (\text{kg/m})$$

and we obtain the objective function:

$$\mu = \frac{n \cdot \left(g_{1c} + g_{1\ddot{e}} \right)}{B} = \frac{q_{1c} + q_{1\ddot{e}}}{b} \to \min.$$

The depth of tillage by a flat cutting cultivator is usually a given value; it has a decisive influence on the weight of the unit, its draft resistance. In connection with this we need a particularly critical approach to the appointing (taking into account) this parameter, as the agronomic basis for appointing the depth of the working body stroke is often highly subjective, often not economically justified.

5. ANGLE OF SOIL CUTTING BY A TRIHEDRAL WEDGE

In Section 3.1 we introduced the concept of "the cutting angle of the soil" (figure 17, equation (20)). Let us specify this determination. The angle between two lines in space, set by of their canonical equations:

$$\frac{x - x_1}{l_1} = \frac{y - y_1}{m_1} = \frac{z - z_1}{n_1};$$

$$\frac{x - x_2}{l_2} = \frac{y - y_2}{m_2} = \frac{z - z_2}{n_2},$$

is determined by the formula

$$\cos(180 - \chi) = \frac{l_1 \cdot l_2 + m_1 \cdot m_2 + n_1 \cdot n_2}{\sqrt{l_1^2 + m_1^2 + n_1^2} \cdot \sqrt{l_2^2 + m_2^2 + n_2^2}}. \tag{40}$$

Suppose that one of the lines will be a trajectory of the soil movement on the wedge (S), and the other – the direction of the wedge movement (OX), then the angle between them χ is a cut angle (fig. 38); $\chi + \rho = 180°$.

Thus, **an angle of cutting of soil** χ – is an angle between the direction of the wedge movement and the direction of the soil movement on the wedge.

Such a definition of the cutting angle of the soil for the first time is introduced by us in the author's certificate SU 1771549.

The equation of the line passing through the point O (0; 0; 0) и C (0; OC; 0) (fig. 38) has the form

$$\frac{x}{l_2} = \frac{y}{m_2} = \frac{z}{n_2},$$

where $l_2 = 0$; $m_2 = OC$; $n_2 = 0$.

The direct line S passes through the points C $(0; OC; 0)$ and D $(OC \cdot \mathrm{tg}\eta_{\check{A}}; OC \cdot \mathrm{tg}\eta_{\check{A}} \cdot \mathrm{tg}\eta_{\check{A}}; 0)$ has the direction cosines $(\cos\alpha_S; \cos\beta_S; \cos\gamma_S)$:

$$\begin{cases} \cos\alpha_S = \cos\eta \cdot \sin\gamma - \cos\varepsilon \cdot \cos\gamma \cdot \sin\eta, \\ \cos\beta_S = \cos\eta\cos\gamma + \cos\varepsilon \cdot \sin\gamma \cdot \sin\eta, \\ \cos\gamma_S = \sin\varepsilon \cdot \sin\eta. \end{cases}$$

We substitute in the equation (40) the values of the direction cosines:

$$\cos(180 - \chi) = \cos\chi \frac{1}{\mathrm{tg}^2\eta_{\check{A}} + 1 + \mathrm{tg}^2\eta_{\check{A}} \cdot \mathrm{tg}^2\eta_{\hat{A}}}.$$

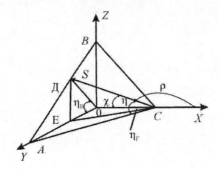

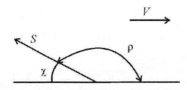

Fig. 38. Cutting angle of the soil by a triangular wedge

After substituting $\text{tg}\eta_\Gamma$ и $\text{tg}\eta_B$ (formulas 25, 26):

$$\text{tg}\eta_{\check{A}} = \frac{\text{tg}\gamma - \cos\varepsilon \cdot \text{tg}\eta}{1 + \cos\varepsilon \cdot \text{tg}\gamma \cdot \text{tg}\eta}; \ \text{tg}\eta_{\check{A}} = \frac{\text{tg}\eta \cdot \sin\varepsilon}{\cos\gamma\left(\text{tg}\gamma - \cos\varepsilon \cdot \text{tg}\eta\right)}.$$

We get:

$$\cos\chi = \cos\gamma \cdot \cos\eta + \cos\varepsilon \cdot \sin\gamma \cdot \sin\eta.$$

Since $\text{tg}\eta = \text{tg}\gamma \cdot \cos\varepsilon$, then

$$\cos\chi = \sqrt{\cos^2\gamma + \cos^2\varepsilon \cdot \sin^2\gamma}. \tag{41}$$

We write three more equations establishing a connection with the cutting angle, the direction of the soil movement on the wedge and the wedge angles:

$$\begin{cases} \cos\chi \cdot \sin\eta = \cos\varepsilon \cdot \sin\gamma, \\ \cos\chi \cdot \cos\eta = \cos\gamma, \\ \sin\chi = \sin\gamma \cdot \sin\varepsilon. \end{cases} \tag{42}$$

The analysis of the equation (42) shows that when $\gamma = 90°$ $\chi = \varepsilon$.

The values of the angle χ for the most used in the working bodies of tillers angles γ, ε are shown in figure 39.

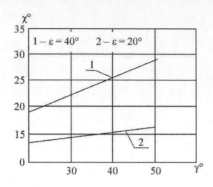

Fig. 39. Dependence of the cutting angle of the soil χ on the wedge parameters γ, ε

Wedge with a variable angle of cut

On the basis of numerous studies [4, 8, 20, 28, 30, 46, 51, 55], it was noted that to obtain high-quality soil treatment with minimal expenses, the working bodies of tillers with specific parameters which appropriate to the kind and condition of the treated soil are necessary.

On the basis of years of research O.V. Vernyaev [19] came to the conclusion that in the conditions of different humidity, density and infestation of soil the most qualitative processing is carried out by the working bodies, in which the angle ε has different meanings. To the same conclusion comes M.D. Podskrebko. [36] in the study of the optimal setting angles of the ploughshare to the bottom of the furrow.

The research by V.A.Zheligovskiy, G.N. Sineokov, P.N. Burchenko, V.I. Vinogradov and their disciples [16, 17, 20, 21, 53] ascertains an objective law that associates the index of crumbling with the degree of a stress condition for an elastic layer of the soil.

On the basis of the analysis of literature, the main schemes of changing the angle ε and the performed research, we have hypothesized about the impact of a variable angle of cutting on the improvement of the soil crumbling. Calculations show that the working bodies with the variable angle of cutting creates more stress in the formation of soil due to the deformation of bending and torsion.

We have investigated the possibility of obtaining a variable angle of cutting on the subsurface tillage working bodies by changing the parameters of the wedge.

By changing the opening angle of the working body 2γ by the use of other forms of ploughshares (concave or convex) (fig. 40, c), we obtain a more rapid wear of plowshares, increasing traction resistance of the working bodies and the width of the furrows, reducing the safety of stubble, etc.

The mechanism of changing the angle 2γ (fig. 40 b) in the process does not also give a positive result: first, it is impossible to change the angle 2γ in a wide range, because it is bounded, on the one hand, by the performance of the condition of cutting with a slip and, on the other hand, − by the productivity of a specific tool and optimal load of the tractor. That is, the increase in the angle 2γ leads to the fact that the soil will not slip on the working surface of ploughshares; there will be a cluster of weed roots on plowshares and soil collecting. In this case, there will not be the quality of soil treatment. Decreasing the angle 2γ will result in increasing the number of racks of the working bodies and, as a consequence, increasing the metal content of soil processing. In addition, there are problems with the regulatory capacity of the unit and the optimal load of the tractor (the decrease of the coverage width of working bodies and instruments in general, the increase of the weight of tools, the decrease of mobility, and so on). All this leads to a failure of agro requirements, a significant rise in the processing of the soil.

A brief presented analysis indicates that the obtaining of the variable angle of cutting on the flat cutting working bodies by the change in the angle 2γ is ineffective.

As already mentioned, for the quality processing ,each particular type of soil should correspond to a working body with specific parameters of the wedge, in particular, with a specific angle of setting of the share to the furrow bottom ε (fig. 41 a). With the given angle ε qualitative processing of a particular type of soil is observed.

However, nowadays there are more than 70 different types of soil in the world. In this regard, a large variety of trademarks of manufactured tillage working bodies will complicate the conduct of tillage, and its high cost will increase significantly.

During making experiments, especially on over dried solid soils, it was noticed that flat cutting working organs (fig. 41 a) having a constant angle ε along a share length, sheared large pieces of soil, the so-called "suitcases", in this connection, in the given fields the additional tillage to break the "suitcases" and leveling the surface of the field was necessary.

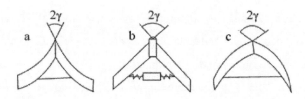

Fig. 40. Change schemes of a cutting angle (χ) by an opening angle (2γ) of the flat cutting foot: a − working body CPG with a regulatory mechanism of an opening angle, b − a working body KPG with convex shares

It was also noted that each individual setting angle of a plowshare corner to the bottom of the furrow ε corresponds to certain size pieces of treated soil, i.e., the size of a piece of treated soil depended on and was determined by the size of the angle ε of the working body.

To eliminate the problem of "suitcases" in cultivation, several researchers [20, 29] offered toothed working bodies (fig. 41 b), and flat cutting working bodies with extra blades fixed on the shares (fig. 41). A toothed working body crumbles the layer of soil very well, but in the process it becomes clogged with plant residues and root system of plants, tooth wear is uneven along the length of plowshares and this working body is complicated to manufacture.

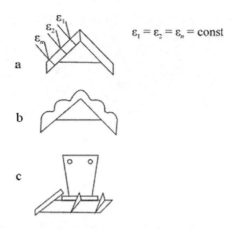

$\varepsilon_1 = \varepsilon_2 = \varepsilon_n = const$

Fig. 41. Various schemes to change the cutting angle χ by the angle ε: a – a serial working body CPG, b – a toothed working body CPG with additional blades

The flat cutting working bodies with additional blades could not solve the problem of "suitcases", because the chipping of the soil layer was done by a plowshare with a constant setting angle to the bottom of the furrow ε, and extra blades raised the soil "suitcases" higher and helped them overturn.

However, getting the variable cutting angle on the flat cutting working bodies by changing the angle ε is preferable, because it enables to achieve the specified quality crumbling of the soil layer.

In the Chelyabinsk State Agroengineering Academy the working bodies of the flat cutting and deep loosening cultivator with an increasing and decreasing cutting angle from the sock to the heel of the share (fig. 42, 43) were proposed and investigated. The working bodies with variable angles of cut (χ) along the length of plowshares at the same energy costs provided the crumbling of the soil as compared with serial 20 ... 50% better. These working bodies crumble the layer well without cutting away large chunks of soil, contribute to a more sustainable stroke of tools in the depth of treatment (fig. 44).

Fig. 42. Working body of the flat cutting and deep loosening cultivator with a variable angle of cut (the cutting angle increases from the sock to the heel)

Fig. 43. Working body of the flat cutting and deep loosening cultivator with a variable angle of cut (the cutting angle decreases from the sock to the heel)

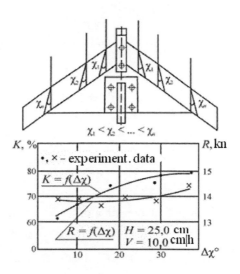

$$\chi_1 < \chi_2 < ... < \chi_n$$

Fig. 44.Dependence of crumbling the soil layer (K) and a traction resistance (R) on the intensity of the change of a cutting angle of the soil ($\Delta\chi$) of the working body CPG with an increasing cutting angle from the sock to the heel of the plowshare:

$$\Delta\chi = \chi_n - \chi_0 \text{ –the intensity of the change of a soil cutting angle,}$$

where χ_0 – the minimum angle of cut

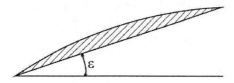

Fig. 45. Characteristic shape of the soil sticking to the flat wedge

It has been hypothesized about the possibility of the existence of such a form of a cultivator foot in which there is no sticking. The hypothesis is based on the following arguments: during the foot motion in the soil the foot surface and the soil interact with the normal to the surface of the wedge (fig. 46). The component N_z of the wedge reaction is the force that moves the ground up.

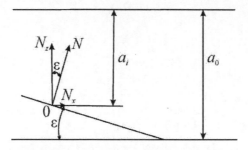

Fig. 46. Vertical component of the reaction of the wedge in contact with soil

Assuming that the reaction N in the soil of any moisture and any mechanical structure depends only on the depth of the working body stroke, we can write:

$$N_i = ka_i,$$

where k – the characteristic of this type of soil by resistivity, N / m ^{2};

a_i – the depth at which N is fixed.

$$N_z = N \cdot \cos\varepsilon = ka_i \cos\varepsilon.$$

From this expression it is clear that the driving force N_z is reduced with the decreasing depth. This leads to sticking of the working surface. In order not to have sticking, you need to have such a work surface, on which at any point the condition: N_z = const is observed, i.e.

$$N_z = N \cdot \cos\varepsilon = ka_i \cos\varepsilon = \text{const}.$$

To reach this condition is only one way – to make the angle ε variable in height, i.e.

$$ka_0 \cos\varepsilon_0 = ka_i \cos\varepsilon_i, \tag{43}$$

where a_0 –the value of the stroke at the base of the wedge;

a_i – the value of the stroke depth in the i-th point of the wedge;

ε_0 – the angle of the wedge setting to the furrow bottom at the depth a_0;

ε_i – the angle of the wedge setting to the furrow bottom at the depth a_i.

From the equation (43) we find the angle of the wedge, in which the condition $N_z = $ const is observed:

$$\cos\varepsilon_i = \frac{a_0}{a_i}\cos\varepsilon_0,$$

from which we obtain the equation for the construction of a surface provided $N_z = $ const:

$$\varepsilon_i = \arccos\frac{a_0}{a_i}\cos\varepsilon_0. \tag{44}$$

Figure 47 shows the calculated shape of the working surface of the wedge with $\varepsilon_0 = 30°$, $a_0 = 20$ cm.

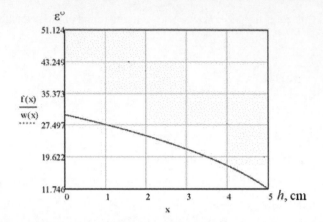

Fig. 47. Estimated value of the angle ε in the height of a triangular wedge: γ0 =
40 °, ε0 = 30 °, a = 20 cm

In the case of the triangular wedge the function of the angle ε performs the
cutting angle χ:

$$\sin \chi = \sin \varepsilon \cdot \sin \gamma ;$$

$$\chi_i = \arccos \frac{a_0}{a_i} \cos \chi_0 . \qquad (45)$$

If the working parts are made with a constant angle of cut along the length of a
foot ($\chi = \chi_0 = \chi_i$ = const), then

$$\sin \varepsilon_i = \frac{\sin \chi_i}{\sin \gamma} ; \qquad (46)$$

$$\sin \chi_0 = \sin \varepsilon_0 \cdot \sin \gamma_0 . \qquad (47)$$

A foot profile can be constructed using the ratio

$$h = d \operatorname{tg} \varepsilon_i ,$$

68

where h – the value of the height of the wedge above the horizontal dimension d.

Figure 48 shows the calculated profile of the foot cultivator with the initial values of angles: $\gamma_0 = 45°$, $\varepsilon_0 = 35°$.

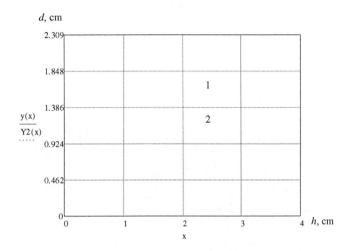

Fig. 48. Calculated profile of a cultivator foot with a variable angle of the cutting in the height of the wedge $\varepsilon 0 = 30\,°$, $a0 = 20$ cm (1 – a flat wedge, 2 – curved wedge)

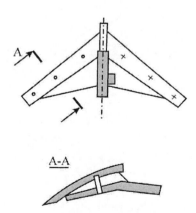

Fig. 49.Ffoot profile of an experimental cultivator

6. EXPERIMENTAL DETERMINATION OF THE TRAJECTORY OF LAYER MOTION ALONG A WEDGE

The essence of a method of determining the angle η during the soil movement on the wedge (a.s. SU 1771549) is as follows: during the motion of the soil on a trihedral wedge along the path defined by the angle η, there is a displacement of the soil in the horizontal plane by the amount of the OE (fig. 50). This offset value can be calculated and experimentally measured.

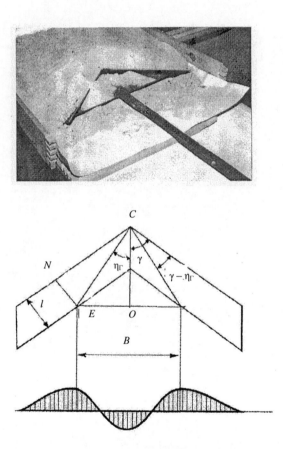

Fig. 50. Scheme to determine the width of a furrow B and the corner η_Γ

$$OE = CE \cdot \sin \eta_{\tilde{A}};$$

$$\sin\left(\gamma - \eta_{\tilde{A}}\right) = \frac{NE}{CE},$$

where NE – the horizontal projection of the width of the share, along which the soil moves.

If l – the share width (wedge), then $NE = l \cdot \cos\varepsilon$. So, we can write

$$CE = \frac{l \cdot \cos\varepsilon}{\sin\left(\gamma - \eta_{\tilde{A}}\right)}.$$

Consequently,

$$OE = \frac{l \cdot \cos\varepsilon \cdot \sin\eta_{\tilde{A}}}{\sin\left(\gamma - \eta_{\tilde{A}}\right)}.$$

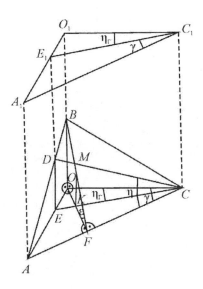

Fig. 51. Determining the direction of the relative trajectory of the soil movement on the surface of the working body

Since

$$\hat{A} = 2 \cdot OE,$$

where B – the width of an open furrow along the ridges, then

$$B = 2l \cdot \frac{\cos\varepsilon \cdot \sin\eta_{\hat{A}}}{\sin(\gamma - \eta_{\hat{A}})}. \qquad (48)$$

The experimental values of the open furrow width (B) at different angles ε and γ are shown in figure 52.

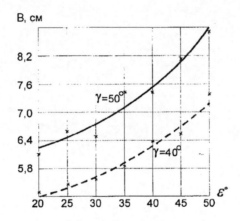

Fig. 52. Width of an open furrow B, depending on the parameters of the wedge γ °, ε °

The horizontal projection of the angle η – the angle ηG angle – can be measured experimentally (a.s.SU 1771549). The agreement between the experimental measurements can be performed as follows. The equation (48) can be written as

$$\frac{\sin(\gamma - \eta_\varepsilon)}{\sin\eta_\varepsilon} = \frac{2l}{B} \cdot \cos\varepsilon$$

or

$$\frac{\sin\gamma\cdot\cos\eta_{\bar{A}} - \cos\gamma\cdot\sin\eta_{\bar{A}}}{\sin\eta_{\bar{A}}} = \frac{2l}{B}\cdot\cos\varepsilon;$$

$$\sin\gamma\cdot\operatorname{ctg}\eta_{\bar{A}} = \frac{2l}{B}\cdot\cos\varepsilon + \cos\gamma;$$

$$\operatorname{ctg}\eta_{\bar{A}} = \frac{2l\cdot\cos\varepsilon}{B\cdot\sin\gamma} + \operatorname{ctg}\gamma. \tag{49}$$

The experimental values of the angle η_Γ at different values of angles γ and ε are shown in figure 53. Since the angle η is directly very difficult to measure, its value can be found on the magnitude of its projection on a horizontal plane −the corner η_Γ.

The equation (25) can be written as

$$\operatorname{tg}\eta_{\bar{A}} + \operatorname{tg}\eta_{\bar{A}}\cos\varepsilon\operatorname{tg}\gamma\operatorname{tg}\eta = \operatorname{tg}\gamma - \cos\varepsilon\operatorname{tg}\eta.$$

Hence

$$\operatorname{tg}\eta = \frac{\operatorname{tg}\gamma - \operatorname{tg}\eta_{\bar{A}}}{\cos\varepsilon(\operatorname{tg}\gamma\operatorname{tg}\eta_{\bar{A}} + 1)}. \tag{50}$$

The experimental values of the angle η, obtained by applying to the recalculations of experimental values of the angle η_Γ are shown in figure 54.

Changing the values of the angles γ, ε, and conducting quite a large number of experiments for the measurement of the value B one can calculate fairly accurately the value of the angles η_Γ and η.

It should be noted that the error of determination of the angle η with this method greatly increases with the increase of the share stroke depth. For a> 10h on the surface of the soil the movement is not almost fixed (h − the height of the wedge, a −

the depth of the wedge stroke in the soil). Therefore, we proposed other methods for determining the angle η (a.s. SU 1771549, 1813316, RF patent 2193796).

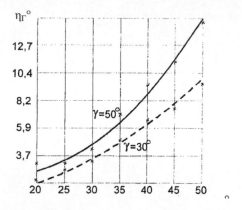

Fig. 53. Dependence of the horizontal projection of the angle (η_r) on the parameters of the wedge ε, γ

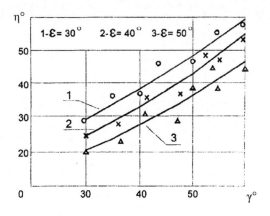

Fig. 54.Dependence of the angle η ° on the wedge parameters ε °, γ °: Δ − experimental data by V.K. Sharshakov; × − experimental data by L.D. Turaev;
○ − experimental data by P.G. Svechnikov

The experimental studies to determine the angle η by different methods showed the approximate equivalence, and accuracy of the results. The experimental results clearly demonstrate the validity of the equation (19).

Fig. 55. Experimental installation to determine the width of an open furrow B and angle $η_Γ$

The issue of the impact of the speed of the wedge on the angle η continues to be controversial . L.V. Gyachev [26] states that the angle of the entry of sand onto the wedge plane does not depend on the wedge setting angle towards the furrow bottom and the speed (V = 1 m/s, h = 7 ... 8 cm). V.K. Sharshak [57] notes that with the increasing of the speed of the working body, there is some tendency to increase the angle of the layer entry onto the share.

A special laboratory installation, made in the Chelyabinsk State Agroengineering Academy by V.V. Kulagin, enables to simulate the speed of the movement of soil (sand) in the range 1.0 ... 3.5 m/s due to the height of the fall of the material, to change within any range the setting angles of the working body (γ and ε), to set the coefficients of friction / slip of the soil due to different material surfaces of sliding and test environment (RF Patent 2193796). Despite the simplicity of the measuring instruments used, a high accuracy of determining the angle η (error not exceeding 3%) due to the availability of the process for the direct measurement and the ability to manage measurement is achieved. With the help of the proposed devices it has been established experimentally that on the trajectory of the soil on the wedge, the movement speed and the coefficient of the soil friction on the surface of the wedge do not have a significant effect (fig. 56, 57).

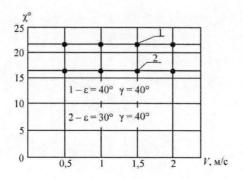

Fig. 56. Dependence of the cutting angle of the soil by a flat trihedral wedge (χ) on the speed of movement (V)

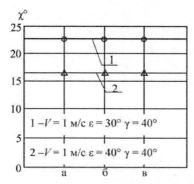

Fig. 57. Cutting angle of soil by a flat trihedral wedge depending on the material of the sliding surface (soil, clay-humus): a − polished steel; b − polyethylene coating, c − gray cast iron

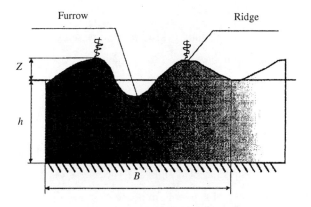

Fig. 58. Formation of characteristic irregularities on the surface of the field (cross-section)

The study of this question allowed us to propose the following scheme of the formation of irregularities and an open furrow by flat cutting and deep loosening cultivators:

1. Characteristic elevations are formed from the action of a rack similarly like the elevations formed during the vertical movement of a vertical rod in the sand, or other environment. At the same time, an initial open furrow is being formed.

2 .An open furrow increases during the layer movement on the wedge with an angle η to the blade.

The effect of a rack on the size of irregularities of a field and an open furrow was investigated by us in a laboratory setting in a soil channel (ChSAA).

During the experiment, we changed the following characteristics: the rack thickness δ = 30 ... 60 mm, the rack stroke depth h = 8 ... 25 cm, the speed of movement V = 0.3 ... 2.2 m/s. The hardness and moisture of soil had values, respectively, 2.3 ... 2.9 ... 15 MPa and 18%.

7. INFLUENCE OF A CONSTRUCTIVE DESIGN OF WORKING BODIES ON A TECHNOLOGICAL PROCESS

The processes of soil movement on flat working bodies, which we have studied, in their pure forms are very rare practically. The fact is that the working body should in any way be attached to the frame of a tillage tool, and the nature of the attachment can have a significant impact on the movement of soil on the working body.

However, this phenomenon has not yet been taken into account by researchers of working bodies of tillers. It is well known that after the passage of flat cutting and deep loosening cultivators on the surface of a field the roughness of specific forms that adversely affects the performance of subsequent processing steps is left (fig. 58).

The attempts to explain these experimental data only by the motion of the soil on the working body usually lead to incorrect theoretical models. Thus, the work [27] states that the angle η is more than the angle γ, thereby two opposing flows of soil are created which, by the interaction between themselves and a rack of a working body form typical unevenness. To the same conclusion comes N.V. Ivanov [28]. Many writers are limited to a statement of the fact of the influence of a rack of the working body on quality indicators.

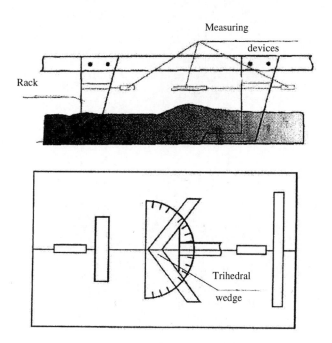

Fig. 59. Scheme of a laboratory apparatus for determining the width of an open furrow from the impact of a wedge and a rack

Experiments have shown that the increase in the movement speed, the rack thickness and the depth of the rack stroke contribute to increasing the size of the field irregularities and the width of the open furrow.

The experimental results showed that the formation of the field irregularities and the open furrow of the field is carried out by the scheme shown in figure 60.

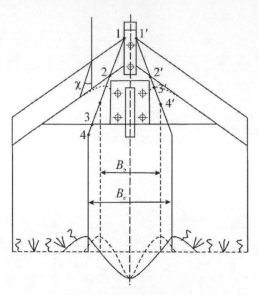

Fig. 60. The scheme of the formation the field irregularities and the open furrow y the flat cutting working bodies

A particle of soil, moving along a share at the angle η to the blade from point 1 to point 2 goes further, moving on a shoe, or, rather, on the ground, lying on the shoe, goes to point 3 and then falls to the ground from the shoe − point 4. Thus, the soil particlesturn out to be displaced from the longitudinal axis of symmetry of the working body by the distance B/2, which is determined by the equation

$$B = 2\left(d + l_{\text{л}} + l_6 + l_V\right), \qquad (51)$$

where d − half the width of a bit;

$l_{\text{л}}$ − the displacement caused by the movement of soil on a share;

l_6 − the displacement caused by the movement of soil on a shoe;

l_V − the displacement caused by the coming of from a shoe at the speed V.

One can consider with a margin of error the movement of soil on the ploughshare and the shoe to be in the direction toward $η_г$ − the horizontal component of the angle η:

$$l = l_{\text{л}} + l_6 = mn \cdot \text{tg}\eta_{\Gamma},$$

where mn – the linear size of the ploughshare and the shoe in the direction of motion;

$$l_V = V \cdot \sqrt{\frac{2H}{g}} \cdot \sin \eta_{\Gamma}$$ – the shift caused by the speed of movement of the tools;

V – the speed of the movement of the tool;

H – the height of the wedge;

g – the acceleration of gravity.

The shift l is caused by the direction of the movement trajectory of the soil on the wedge, as the share and the shoe in the work process form a single wedge.

From the equation (46) follows a very important practical conclusion: the width of the furrow can be significantly reduced by removing the shoe. In order to test this hypothesis, a series of field tests of serial working bodies and working bodies that practically had no shoes (fig. 61).

Fig. 61. Recommended (experimental) working body of a flat cutting and deep loosening cultivator

The test results showed that the working bodies without the shoe contribute to the significantly lower width of the open furrow (fig. 62).

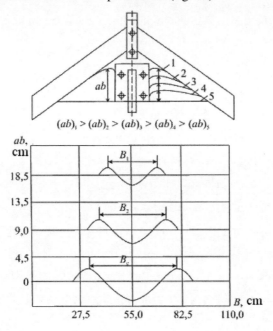

Fig. 62. Dependence of the width of the open furrow *(B)* on the shoe protrusion size *(ab)*

Conclusion

Scholars and practitioners around the world are constantly looking for such technologies, which would allow increasing the productivity of the soil and animals while reducing energy consumption per unit of production. Similar objectives are also in front of other sectors of the economy. Therefore, in today's world very dynamic replacement of structural materials for lighter, stronger, improved ones takes place, the technology of the manufacture of parts becomes more perfect, it is possible to make the parts more complex in configuration, light and reliable. This is what causes the need for continuous updating of structures of agricultural machinery.

What yesterday was not possible because of lack of necessary materials or manufacturing techniques for producing working bodies today may be an ordinary affair. At the same time while changing the technologies of cultivating crops, the performed operations, the nature of the interaction of the working bodies with the environment change. Scientific and engineering development of the working bodies is necessary.

In mechanics, as in any branch of natural science, the study of a real object begins with the selection of a calculation scheme – a model (by some researchers a calculation scheme is called a mathematical model that is not quite correct as a mathematical model – this is only an equation).

The design scheme is an image of a real object without inessential singularities, details, interactions, etc. If a new tool has the same design scheme with the old tool, the creation of a new tool becomes a routine job and can be automated. In this case, the research part is almost completely eliminated. In most cases, the process of creating a new tool begins with a long search of a suitable (adequate) design scheme.

At the core of creating a design scheme of machines and tools there are three basic laws of mechanics, for the first time in this form formulated by Isaac Newton.

The knowledge of and proper application of Newton's laws allow theoretically, at the design stage to determine the force of interaction of contacting bodies and on their basis to find the parameters of agricultural machinery.

It is especially important in creating machines to predict their behavior during operation. Every machine is designed to perform the useful work, so during the creation of agricultural machinery the quality of technological operations is always in the first place.

In the process of doing useful work the working bodies of tools interact with the soil, plants and fertilizers. This interaction is determined by the type of a technological operation.

A wedge is in the heart of tillage working bodies. The discussed in this paper of the questions of the theory of the tillage wedge and the examples of its application in practice revealed internal dependencies and relationships that have a major impact on the quality of the process of tillage.

The investigation of the basic features of the interaction of tools with the environment enables to determine the forces arising in this case. The forces acting on the working bodies are the basis of all calculations of a tool and determine the type of traction means. The values of forces enable further to choose the material for the manufacture of all the elements of a tool, make the necessary strength calculations, determine the cost of energy to perform work, the efficiency of a tool and technical and economic parameters of the soil-cultivating units.

The obtained theoretical laws of the interaction of the soil tillage wedge enable to create new designs and upgrade existing tillage working bodies and ensure the specified quality of tillage, which ultimately leads to receiving the agricultural product in the desired amount, the required quality, with minimal compared with existing technologies and hardware costs.

Recently, fewer and fewer scientific papers on the theory of the tillage wedge have appeared. There are a lot of reasons: first, it is hard work that requires a systematic knowledge of the set of branches of science, and secondly, the results of the theoretical investigations are not accepted by people at once. History shows that it takes a long time for theoretical principles to become the basis for fundamentally new technical solutions and technologies, and thirdly, the time in which we live requires more practical solutions that bring an immediate result rather than theoretical refinements that maybe sometime in the future will be the basis for revolutionary change. In this connection, the work of a theoretical plan, we believe, must be maintained at the state level as well.

We express the hope that the presented piece will be useful to the work of scholars and practitioners, design and engineering organizations, researchers, engaged in tilling the soil.

References

1. Bakhtin P.U. Investigation of physical, mechanical and technological properties of the major soil types of the USSR. M.: Spike, 1969. 268 p.

2. Bakhtin P.U., Volotskaya V.I., Nikolaev I.I. Coefficient of sliding friction metal-soil of main soils of the USSR / / Tractors and farm machinery. 1964. №6. P. 31-33.

3. Blednykh V.V. Interaction of the blade of weeding feet of a cultivator with weeds / / Tractors and farm machinery. 1979. № 4. p. 17-20.

4. Blednykh V.V. Interaction of the soil layer with the surface of the wedge / / Improvement of the methods of use and maintenance of farm machinery: col. of scien. pap. / ChIMEA. Chelyabinsk, 1984. p. 36-40.

5. Blednykh V.V. Kinematics of moldboard plowing / / Col. of scien. pap. / ChIMEA. Chelyabinsk, 1983. p. 14-24.

6. Blednykh V.V. Basic laws of soil movement on a trihedral wedge / / Dynamics of tillage machines and working parts for tillage: Col. of scien. pap. / ChIMEA. Chelyabinsk, 1982. p. 4-14.

7. Blednykh V.V. Main laws of force interaction of a trihedral wedge with soil / / Advances in science and technology of APC. 2008. № 8. p. 33-36.

8. Blednykh V.V. Improving the working bodies of tillers based on mathematical modeling of technological processes: dis. ... Doctor Techn. Science. L., 1989. 230 p.

9. Blednykh V.V. Technological basis for determining the parameters of working bodies of tillers / /VISKhOM. Ser. "Agricultural machinebuilding, machinery, aggregates and units." 1978. Issue 2.P. 18-24.

10. Blednykh V.V. Structure, calculation and design of tools for the treatment of soil. Chelyabinsk: ChSAA, 2010. 214 p.

11. Blednykh V.V., Buryakov A.S. Curved shape of the working surface of a ploughshare for flat cutting tools/ / Materials of the First Zonal Conference of Young Scientists of Northern Kazakhstan. Tzelinograd, 1969. P. 295-296.

12. Blednykh V.V., Buryakov A.S. Justification of the form of a wedge of a flat cutting cultivator: Col. of scien. pap /ChIMEA. Chelyabinsk, 1970. Issue 56. P. 169-172.

13. Blednykh V.V., Levanidov V.V , Svechnikov P.G. Formation process of a layer of soil by a working body / / Mechanization and electrification of agriculture. M.:VO Agropromizdat, 1986.№ 4. P. 18-20.

14. Blednykh V.V., Svechnikov P.G. Working body for subsurface tillage / / Equipment in agriculture. 1984. № 5. P. 54-55.

15. Blednykh V.V., Svechnikov P.G. Working body KPG with a variable angle of cutting / / Mechanization and electrification of agriculture. 1986. № 5. P.55.

16. Burchenko P.N. Mechanical and technological basics of tillage machines of a new generation. M.,: VIM, 2002. 62 p.

17. Burchenko P.N. Mechanical and technological justification for the parameters of tillers of a new generation for the work in an optimal speed range: summary. dis. ... Doctor Techn. Science. M., 1987. 44 p.

18. Vasilenko P.M. Cultivators (design, theory and calculation). Kiev: TsSKhA, 1961. 156 p.

19. Vernyaev O.V. Theory, design and study of an active working member of a cultivator: summary. dis. ... Candidate Techn. Science. Kharkov, 1960. 31 p.

20. Vinogradov V.I., Morozov N.I. Toothed ploughshare / / of tools for primary tillage: materials NTS VISKhOM. M., 1959.Issue. 5. P. 494-530.

21. Vinogradov V.I., Podskrebko M.D. Effects of speed on the amount of normal and tangent forces acting on the surface of a flat wedge/ / Increase of operating speeds of tractors and agricultural machinery / under total. ed. by V.N. Boltinskiy. M., 1963. P. 210-218.

22. Goldshtein M.N. Mechanical properties of soils. 2nd ed., revised. M. Stroyizdat, 1971. 367 p.

23. Goryachkin V.P. Wedge theory: collec of works. Moscow: Spike, 1965.V. 2. P. 382-389.

24. Goryachkin V.P. Theory destruction of soils: collec. of works. Moscow: Kolos, 1965.V. 2. P. 369-381.

25. Gyachev L.P. Theory of plowshare-moldboard surface / / Tr. Azov and Chernomor. IMESKh. Zernograd, 1961.issue 13. 60 p.

26. Zhilkin V.A. Calculations of strength and stiffness of the elements of agricultural machines. Chelyabinsk: ChSAU, 2004. 428 p.

27. Zhuk M.Ya., Rubin V.F. On the resistance of soil to various deformations / / Tillage machine: collec. of works / VISKhOM / ed.by Prof. NV Schuchkin. M., L. Mashgiz, 1940. Issue 3. P. 35-37.

28. Ivanov, N.V. Influence of parameters and the speed of the movement of a working body of a flat cutting cultivator on work indicators / / Works of ChIMEA. Chelyabinsk, 1974. Issue 77. P. 20-22.

29. Kushnariov A.S. Mechanics and technological basics of the process of impact of working bodies of tillage machines and tools on soil: dis. ... Doctor Techn. Science. Melitopol, 1980. 329 p.

30. Listopad G.E., Kashevarov F.M. On deformation of soil by working bodies oftillers / / Reports of VASKhNIL. 1973.№ 10.P. 42-44.

31. Lurie A.B., Lyubimov F.I. Wide coverage tillage machine. L: Machinebuilding 1981. 142 p.

32. Matsepuro V.M. Investigation of laws of the resistance of soils. Minsk, 1967. 210 p.

33. Pavlov A.V., Korabelskiy V.I., Pavlotskiy A. Geometric reasoning of a surface shape that combines the rational cutting of a soil layer with its deformation / / Applied Geometry and Engineering Graphics. 1975. № 9. P. 124-127.

34. Panov I.M., Vetokhin V.I. Physical principles of soil mechanics. Kiev: Harvest, 2008. 89 p.

35. Pigulevskiy M.Kh. Ways and methods of experimental study of soil deformation / / Theory, design and manufacture of agricultural machines. M., 1936. .V. 2. P. 118-141.

36. Podskrebko M.D. Increase of use efficiency of tractor units on primary tillage: dis. ... Doctor Techn. Science. Chelyabinsk, 1975. 320 p.

37.Revut I.B. Physics of soil. L.: Gidrometizdat, 1972. 356 p.

38. Svechnikov P.G. Study of the formation of a layer during the motion of a dihedral wedge in a variety of habitats / / Bulletin of FSBEI HPE "Moscow State Agroengineering University". 2011. Ser. "Agroengineering."Issue 2 (47). P. 53-55.

39. Svechnikov P.G Direction cosines of the line of the layer movement on a trihedral wedge / / Scientific Review. , 2011. №. P. 115-119.

40. Svechnikov P.G. Determination of the trajectory of soil movement by the value of an open furrow / / Scientific Review. , 2011. № 6. P. 119-123.

41. Svechnikov P.G. Optimal profile of a foot of a flat cutting cultivator / / Tractors and farm machinery. , 2012. № 1. P. 40-41.

42. Svechnikov P.G. Flat cutting cultivator with a shaped shoe / / Ural fields. 1985. № 12. P. 17-18 p.

43. Svechnikov P.G. Plough with a variable cutting angle / / Rural farmer. 2012. № 3. P. 11-12.

44. Svechnikov P.G. Working body of a deep loosening cultivator with a variable cutting angle for the tillage of the soil exposed to wind erosion / / Inform. sheet № 417-83. Chelyabinsk: TsNTI, 1983. 2 p.

45. Svechnikov P.G. Working body of a flat cutting and deep loosening cultivator for subsurface tillage / / Ural fields. 1984. №1. P. 60-61.

46. Svechnikov P.G. Results of laboratory study of the formation of a layer on a wedge / / Tilling machines and dynamics of machines: collec. of scien. pap. / ChIMEA. Chelyabinsk, 1983. P. 18-23.

47. Svechnikov P.G. Method and an apparatus for determining the trajectory of soil on the working body KPG / / Machinery and equipment for the village. , 2011. № 12. P. 8-10.

48. Svechnikov P.G. Cutting angle of soil by a trihedral wedge / / Scientific Review. , 2012.№ 1. 3. 123-126.

49. Svechnikov P.G. Form of a dihedral wedge with minimal sticking / / Tractors and farm machinery. , 2011. № 12. P. 24-25.

50. Svechnikov P.G., Blednykh V.V. Effect of a variable cutting angle of a working body of a deep loosening cultivator on the crumbling of soil / / Tilling machines and dynamics of machines: collec. of scien. pap. / ChIMEA. Chelyabinsk, 1986. P. 18-23.

51. Svechnikov P.G. Justification of parameters of a flat cutting foot with a variable cutting angle for deep loosening of soil: dis. ... Candidate Techn. Science. Chelyabinsk, 1984. 217 p.

52. Sineokov G.N. Design of tillers. M., 1965. 311 p.

53. Sineokov G.N., Panov I.M. Theory and calculation of agricultural machinery. Machinebuilding, 1977. 328 p.

54. Soloviev S.P. Laboratory testing of a soil cutting process/ / Works of VIM. M., 1967. V. 43. P. 95-106.

55. Terzaghi K. Theory of soil mechanics / tran. from English. M. Gosstroyizdat, 1961. 142 p.

56. Turbin B.G., Lurie A.B. Agricultural machinery. Theory and technological calculation. Ed. 2nd, revised. and add. L: Mechanical Engineering, 1967. 583 p.

57. Sharshak V.K. Fundamentals of the theory of plow working bodies of meliorative tillers: Dis. ... Doctor Techn. Science. Novocherkassk, 1980. 368 p.

V.V. Blednykh – Academician of RAAS, Doctor of Technical Sciences, Professor, Senior Researcher at the Research Department of FSBEI HPE "Chelyabinsk State Agroengineering Academy".

P.G. Svechnikov – Candidate of Technical Sciences, Associate Professor, Associate Professor of the department "Tillage, sowing machines and agriculture" of FSBEI HPE "Chelyabinsk State Agroengineering Academy".